Econophysics

Many thanks to Constantin MANEA
(coordinator for the harmonization of the multidisciplinary scientific language)

Econophysics

Background and Applications in Economics, Finance, and Sociophysics

Edited by
Gheorghe Săvoiu

AMSTERDAM • BOSTON • HEIDELBERG • LONDON
NEW YORK • OXFORD • PARIS • SAN DIEGO
SAN FRANCISCO • SINGAPORE • SYDNEY • TOKYO
Academic Press is an imprint of Elsevier

Academic Press is an imprint of Elsevier
The Boulevard, Langford Lane, Kidlington, Oxford, OX5 1GB, UK
225 Wyman Street, Waltham, MA 02451, USA

First published 2013

British Library Cataloguing-in-Publication Data
A catalogue record for this book is available from the British Library

Library of Congress Cataloging-in-Publication Data
A catalog record for this book is available from the Library of Congress

ISBN: 978-0-12-404626-9

For information on all Academic Press publications
visit our website at store.elsevier.com

This book has been manufactured using Print On Demand technology. Each copy is produced
to order and is limited to black ink. The online version of this book will show color figures
where appropriate.

Working together to grow
libraries in developing countries

www.elsevier.com | www.bookaid.org | www.sabre.org

ELSEVIER   BOOK AID
International   Sabre Foundation

CONTENTS

LIST OF CONTRIBUTORS

Constantin Andronache
Boston College, Chestnut Hill, MA, USA

Radu Chişleag
Faculty of Applied Sciences, Bucharest Polytechnic University, Romania

Aretina-Magdalena David-Pearson
Faculty of Applied Sciences, Bucharest Polytechnic University, Romania

Mircea Gligor
National College "Roman Vodă", Roman, Neamţ, Romania

Ion Iorga Simăn
University of Piteşti, Romania

Gheorghe Săvoiu
University of Piteşti, Romania

PART I

General Background of Econophysics

CHAPTER 1

History and Role of Econophysics in Scientific Research

Gheorghe Săvoiu and Ion Iorga Simăn
University of Piteşti, Faculty of Economics and Faculty of Sciences, Romania

1.1 INTRODUCTION

"Obviously, you can't predict the future, but, such research reveals how physicists and economists should compare notes in the future."
*Eugene Stanley (*Dallas Morning News*, 2000)*

Physics has probably had the more dominating effect on the development of formal economic theory. This kind of observer is not going to forecast that there will be ideas developed by economists that will influence physics in the future. However, the relationship between economics and physics that has transpired in the past two decades seems very likely to be a model for the future. And not only because of the complicated development of two different disciplines that are influencing each other through complex theory but also for the other ideas and for all the potential of its new fields of application. The interactions hold not only between economics and physics, as well as between some other disciplines and either of the former, but even between economists and physicists, as well as biologists, sociologists, etc.

1.2 RELATIONS BETWEEN ECONOMICS AND PHYSICS

Contemporary human sciences are more and more distinctive, beginning with psychology, up to cognitive science, from sociology, to economics, from political science, to anthropology, etc. The special humanistic sciences were previously known as moral sciences, and, at the same time, the human sciences have a tradition of drawing analogies with ideas from the natural world and the natural sciences (Mirowski, 1989). There are a great diversity of schools of economic thought (Blaug, 1992): Austrian economists, institutionalists, Marxists, social economists, behavioral economists, chaos theorists, Keynesians and post-Keynesians, neo-Ricardians, agency theorists, the Chicago School, constitutional political economists, public choice theory (the rational choice theory is already at the center of the discipline, in neoclassical equilibrium microeconomics and also macroeconomics).

Physics was born fundamentally in a demonstrative or reductive way of thinking and then, with Newton, also universal. What is fundamental? Can there be unity without fundamentality? The form that unity, especially in physics, takes and should take is a controversial matter that has led to pluralism within the physics community (e.g., *Stanford Encyclopedia of Philosophy*). Physics developed a real universe of labor between theoretical and experimental work. This process has led to disunity within physics of one kind but also to a more complex set of interactions within physics and, as it has become a fundamental science, with other disciplines such as engineering, economics, and management. A distinctive feature of the economic sciences is that, while they share with physics the descriptive and explanatory application of mathematical statistics—in population and probabilistic interpretations—they seem to lack strict and universal laws of the sort "recognized" in physics (Stanley, 1999).

Economics is oriented toward choice, risk management, and decision-making problems but involves some general aspects that are of particular interest for understanding the real phenomena or processes:

- Economics deals only with a certain aspect of reality, often of an optimum manner, in which man employs scarce resources.

- Economics always encourages the application of quantitative and formal methods, to gain intellectual legitimacy associated with the virtues of precision and objectivity.
- Economic aggregate subjects are somehow made from the same simple units that are individuals: households, financial or nonfinancial corporations and agencies, labor and markets (Gordon, 2000).
- Economic systems are increasing day by day, in the human and natural environments like physical, biological, and social types.

Physics can contribute more or less unexpectedly to understanding economic problems, processes, or decisions:

- With its methodology, that can be described as analytical.
- With its solutions of decomposing a system into its parts and its final manner of understanding known as *"the whole is greater than the sum of its parts."* (the Gestalt phenomenon).
- With its scale of measurement or its quantitative point of view relevant in describing the qualities of an economic system or phenomenon (Gopikrishnan and Stanley, 2003).
- With its specific way to view and to relate in terms the parts of the universe that must be studied within the great hierarchy of the structure of reality: from a microscale perspective to a macroscale perspective, dealing with the two greatest extremes (nuclear physics at the subatomic particle level and astrophysics at the cosmic and universal level of aggregation, from disciplines like chemistry, molecular biology, organic biology, psychology, to economics, political science and sociology, ecology, climatology and geology, and finally to astrophysics).
- With its contribution to establishing equations that simplify and methods that describe phenomena such as production, market, migration, traffic, transportation, and the financial world (Stauffer, 2000).
- Physics seems to have been reluctant to work in areas where data sets are short and unreliable—but this characterizes a great deal of data in the social sciences (*Stanford Encyclopedia of Philosophy*).
- Physics has been motivated from the examination of unity in specific sciences, and so it has been defined as the unifying and unified science (Stanley et al., 1996).

Physics, more than any other scientific discipline, has had a tradition of thought experiments linked to conceptual revolutions. The most interesting framework of social science is the rational choice theory, with formal tools and constraints to characterize individuals' rationality and conditions of aggregativity. This form of reductionism has been applied to the study of social behavior. Theories and models of social behavior are reduced to those of economic behavior, most notably by Becker (1976). It is also a key to the reductive projects about the relation of macroeconomics to microeconomics. Finally, economic methods and models are related to physics (let us remember, for instance, Adam Smith's Newtonianism and Maxwell's application of statistical methods first developed in Quetelet's statistical sociology or "social physics" and of accounting concepts). But to describe all the methods and solutions would be impossible without a deeper understanding of these economics systems having an enormous degree of complexity, understanding offered by applying modern techniques of statistical physics (Plerou et al., 2000). And so emerges the field of a new science: econophysics, a science that offers a new chance of identifying qualities through quantities, to describe the basis of the universal laws (from physics to economics).

1.3 RELATIONS BETWEEN ECONOMISTS AND PHYSICISTS

After more than two decades, some critical aspects are relevant to the new cohabitation between scientific researchers of economics and physics, in the field of econophysics:

• It has became already traditional to argue that physicists and economists belong to the distinct categories of physical or natural (*hard*) science and social (*soft*) science.
• There are some arguments that in fact the teaching of microeconomics and macroeconomics as they are currently constituted should cease and be replaced by appropriate courses in mathematics, physics, and some other *harder* sciences (Mc Cauley, 2004).
• It has been demonstrated that in economics there are some interesting areas where the use of experiments is more efficient and where there are a lot of possibilities to be tested experimentally even more than in physics (Lillo et al., 2003).
• The failure of economists to deal properly with certain empirical regularities is a starting point of argument for econophysics.

- The implicit intellectual superiority of mathematics followed by physics (including physical statistics) in general over economics is a source to potential conflicts between the two groups as the econophysics has developed (Mc Cauley, 2006).
- Regular economists are unwilling to deal with data or facts that do not conform to their predictions or theories, which means that in this regard economists are not really scientists.
- Economists who study the same data and models as the physicists have been known to complain that the physicists apply models to data without any real theory at all (only mutual research and effort will overcome these prejudices through communication).
- While economists are mostly conventional general equilibrium theorists who proved rigorous theorems (classical, neoclassical, or orthodox economy), physicists found them to be absurdly empty and a cover for the lack of real science that the economists were doing.
- What seem to be models of physics sometimes came from economics, and the standard models of economics came more often from physics (the general argument of originality was born there, a truly complicated development or a new name for something that has been going on for a long time).
- Many physicists cannot understand even the simplest supply-and-demand model, as developed by Alfred Marshall (Krugman and Venables, 1990).
- Just a few economists recognize and accuse other economists of suffering from physics envy (Mirowski, 1989).

Some physicists have pointed to the lack of predictive power of economic theory and have assimilated it to an impoverished theoretical geometry with insufficient intellectual resources capable of improving its applicability (Rosenberg, 1992).

Other physicists have pointed to the preeminent role of thought experiments in economic thought (Reiss, 2007) or have advocated a methodology based on the epistemological virtues of researchers—honesty (Guala, 2005). Even in contemporary econophysics there are some tough arguments against new science (Gallegati et al., 2006):

- Some econophysicists lack awareness of what has been done in economics and thus sometimes claim a greater degree of originality and innovativeness in their work than is deserved.

- A lot of econophysicists do not use sufficiently rigorous or sophisticated statistical methodology as econometricians.
- Many econophysicists naively believe and search for universal empirical regularities in economics that probably do not exist.
- Some theoretical models of econophysics used to explain empirical phenomena have many difficulties and limits.

There is considerable substance to all these arguments, but it is interesting that there is no arguement with physicists. The solution remains a greater degree of collaboration between economists and econophysicists in order to resolve and avoid these problems, something that these critics probably agree with (Di Matteo and Aste, 2007).

1.4 HISTORY OF ECONOPHYSICS

In his doctoral thesis entitled *Théorie de la speculation*, at the Academy of Paris, on March 29, 1900, Louis Bachelier determined the probability of price changes. Perhaps this was indeed the beginning of econophysics. Bachelier's original proposal of Gaussian distributed price changes was very soon replaced by a lot of alternative models, out of which the most appreciated was a geometric Brownian motion, where the differences of the logarithms of prices are distributed in a Gaussian manner (Mantegna and Stanley, 1997; Mantegna and Stanley, 2000).

The interest of physicists in financial and economic systems has roots that date back to 1936, when Majorana wrote a pioneering paper, published in 1942 and entitled *Il valore delle leggi statistiche nella fisica e nelle scienze sociali*, on the essential analogy between statistical laws in physics and social sciences. Many years later, a statistical physicist, Elliott Montroll, coauthored with W.W. Badger, in 1974, the book *Introduction to Quantitative Aspects of Social Phenomena*.

Since the 1970s, a series of significant changes have taken place in the world of finance that finally will engender the new scientific field of econophysics. One key year was 1973, when currencies began to be traded in financial markets, and the first paper that presented a rational option-pricing formula was published (Fischer and Scholes, 1973), and immediately after the 1990s, a growing number of physicists have attempted to analyze and model financial markets and, more generally, economic systems. "Today physicists regard the application of statistical mechanics to social phenomena as a new and risky venture. Few, it seems, recall how the process originated the other way around, in the days when physical science and social science were the twin siblings of a mechanistic philosophy and when it was not in the least disreputable to invoke the habits of people to explain the habits of inanimate particles" (Ball, 2004).

1.5 SOME DEFINITIONAL ISSUES

Econophysics was, from the very beginning, the application of the principles of physics to the study of financial markets, under the hypothesis that the economic world behaves like a collection of electrons or a group of water molecules that interact with each other, and it has always been considered that the econophysicists, with new tools of statistical physics and the recent breakthroughs in understanding chaotic systems, are making a controversial start at tearing up some perplexing economics and reducing them to a few elegant general principles with the help of some serious mathematics borrowed from the study of disordered materials.

The term *econophysics* was introduced by analogy with similar terms which describe applications of physics to different fields, such as astrophysics, geophysics, and biophysics. Econophysics was first introduced by the prominent theoretical physicist Eugene Stanley in 1995, at the conference on Dynamics of Complex Systems, which was held in Calcutta, later known as Kolkata, as a satellite meeting to the Statphys 19 conference in China (Chakrabarti, 2005; Yakovenko, 2009). The multidisciplinary field of econophysics uses theory of probabilities and mathematical methods developed in statistical physics to study statistical properties of complex economic systems consisting of a large number of complex units or population (firms, families, households, etc.) made of simple units or humans. Particularly important in defining econophysics is the distinct difference between statistical physics and mathematical statistics in its focus, methods, and results (Malkiel, 2001).

Rosario Mantegna and Eugene H. Stanley have proposed the first definition of econophysics as a multidisciplinary field, or "the activities of physicists who are working on economics problems to test a variety of new conceptual approaches deriving from the physical sciences" (Mantegna and Stanley, 2000). "Economics is a pure subject in statistical mechanics," said Stanley in 2000: "It's not the case that one needs to master the field of economics to study this."

It is a sociological definition, based on physicists who are working on economics problems. Why is econophysics an interdisciplinary science and not a multidisciplinary one? Multidisciplinary suggests distinct disciplines in discussion, as with an economist and a physicist talking to each other. Interdisciplinary suggests a narrow specialty created out of elements of each separate discipline, such as a "water economist" who knows some hydrology and economics.

The more usual way to define a multidisciplinary discipline is to do so in terms of the ideas or methods that it deals with, as for example political economy or biophysics. However, transdisciplinary suggests a deeper synthesis of approaches and ideas from the disciplines involved, and this is the term favored by the ecological economists for what they are trying to develop.

Another, more relevant and synthetic definition considers that econophysics is an "interdisciplinary research field applying methods of statistical physics to problems in economics and finance" (Yakovenko, 2009).

Between econophysics and sociophysics, there are some important differences: while the first focuses on the narrower subject of economic behavior of humans, where more quantitative data is available, the second studies a broader range of social issues. But generally speaking, the boundary between econophysics and sociophysics is not sharp (or crisp), and the two fields enjoy a good rapport.

Econophysics is still a new word, even after 17 years, and is used to describe work being done by physicists, in which financial and economic systems are treated as complex systems. Thus, for physicists, studying the economy means studying a wealth of data on a well-defined complex system (Chatterjee, et al., 2005).

The contemporary way to define econophysics is to do so in terms of the ideas that it involves, in effect physicists doing economics with theories from physics; this raises the question of how the two disciplines relate to each other, and it explains interest rates and fluctuations of stock market prices; these theories draw analogies to earthquakes, turbulence, sand piles, fractals, radioactivity, energy states in nuclei, and the composition of elementary particles (Bouchaud et al., 1999).

On the computer, econophysicists have established a niche of their own by making models much simpler than most economists now choose to consider, even using possible connection between financial or economical terms and *critical points* in statistical mechanics, where the response of a physical system to a small external perturbation becomes infinite because all the subparts of the system respond cooperatively, or the concept of "noise," in spite of the fact that some economists even claim that it is an insult to the intelligence of the market to invoke the presence of a noise term. Many different methods and techniques from physics and the other sciences have been explored by econophysicists, sometimes frantically, including chaos theory, neural networks, and pattern recognition. Another interesting and modern definition considers econophysics a scientific approach to quantitative economy using ideas, models, conceptual and computational methods of statistical physics (Stauffer, 2000). In recent years, many of physical theories like theory of turbulence, scaling, random matrix theory, or renormalization group, were successfully applied to economy giving a boost to modern computational techniques of data analysis, risk management, artificial markets, macroeconomy (Burda et al., 2003). And thus econophysics became a regular discipline covering a large spectrum of problems of modern economy.

A broad definition of econophysics describes it as a new area developed recently by the cooperation between economists, mathematicians, and physicists, which applies ideas, methods, and models in statistical physics and complexity to analyze data from economical phenomena (Wang et al., 2004).

Econophysics is actually nothing more than the combination of the words physics and economics, a link between the two completely separate disciplines that lies within the characteristic behavior exhibited by financial markets similar to other known physical systems. The aim of econophysics is to describe and realize models of the universal behaviors of a market, as an open system, where new external information is mixed with new investments, rather like energy/particle inputs in quantum physics (Jimenez and Moya, 2005). There are some different types of econophysics, too: an experimental or observational type, trying to analyze real data from real markets and to make sense of them, and a theoretical type trying to find microscopic models which give for some quantities good agreement with the experimental facts (Bertrand Roehner, a theoretical physicist based at the University of Paris). First econophysics models published by physicists in a physics journal were those of Mantegna (1991) and Takayasu et al. (1992), though developed a few years earlier. But a Monte Carlo simulation of a market was already published in 1964 by Stigler from the Chicago economics School (Stigler, 1964). The Nobel laureate of Economics H.M. Markowitz published, with Kim, a model for 1987 about the Wall Street crash, with two types of investors, similar to many later models of physicists (Kim and Markowitz, 1989). After the year 2000, econophysics matured enough to allow generalized applications, this field being called sometimes econo-engineering.

Without being similarly defined, econophysics remains the science that uses models taken especially from statistical physics to describe some economic phenomena, an interdisciplinary research field, applying theories and methods originally developed by physicists in order to solve problems in economics, usually those including uncertainty or stochastic elements and nonlinear dynamics. Basic tools of econophysics are probabilistic and statistical methods often taken from statistical physics. Most econophysics approaches, models, and papers that have been written so far refer to the economic processes including systems with large number of elements such as financial or banking markets, stock markets, incomes, production or product's sales, and individual incomes, where statistical physics methods are mainly applied.

1.6 ECONOPHYSICS AND ROMANIAN ECONOPHYSICISTS

As Mirowski (1989) has demonstrated, in his book *More Heat than Light*, the first two centuries of economic theory was modeled primarily on physics. Economics has drawn some inspiration from physics terminology, especially with regard to force, elasticity, equilibrium, velocity, etc. The development of the energy concept in economics was inspired by Western physics and it had subsequent effect on neoclassical economics or modern orthodox theory. The principles of this construction, Newtonian mechanics, implied that the world is guided by elegant deterministic laws, based on a few principles such as the principle of least action. The history of efforts to apply the entropy concept in economics, by both physicists and economists, has been fraught with difficulties. Attempts by physicists to apply the law of entropy date back to G. Helm in 1887 and L. Winiarski in 1900. For them, gold was sociobiological energy and, in 1941, for the econometrician Harold Davis even utility of money became economic entropy.

Obviously, money simply does not perform an equivalent function in economics to what entropy does in physics. Alfred Lotka was the first statistician who noted that there is no general rule regarding the ratio of energy applied to energy set free, but for sure the first economist to deal with entropy was Georgescu-Roegen (1971). He argued that entropy constitutes the ultimate limit and driving force of the economy through ecological relations and processes but nevertheless dismissed more simplistic applications of the law of entropy to the determination of value in economics and claims, among others, were that an economy faces limits to growth, for which he invoked the Second Law of Thermodynamics (useful energy gets dissipated). The name of Nicholas Georgescu-Roegen has been associated not only with entropy but also with a new way of thinking. The main elements of the Roegenian intellectual matrix are still subordination of the entire theoretical and argumentative construction to the general perspective or to the system, transforming economics in a social science where in this case factual reality is the main validation criterion and where economic processes take place according to the law of entropy. Its opinions lead to major reconsideration of neoclassical economic science.

Today, his work is gaining influence, and his insights are being grafted into the new field of evolutionary economics and to econophysics, which follows the Gibbsian statistical mechanics and the nature of statistical distributions, and this new science considers the mathematical structure as an entropic formulation, depending on a conservative system in a pure exchange model framework. Since econophysics was officially born, Romanian scientific researchers in this multidisciplinary field have published many important papers (Gligor and Ignat, 2001). Among these pioneers, one must necessarily mention Sorin Solomon, Adrian Drăgulescu, Radu Chişleag, Mircea Bulinski, Mircea Gligor, Margareta Ignat, among others. From 2003, when the first book entitled *econophysics* was published, in Romania, by Mircea Gligor and Margareta Ignat, followed by *Investment econophysics*, written by Anca Gheorghiu and Ion Spinulescu, 4 years later in 2007, and up to now, new roundtables and satellite workshops have been dedicated to econophysics, including even summer schools of *econophysics and Complexity*—its third and fourth editions were held in 2007 and 2008, respectively.

Since 2008, some of the pioneers of Romanian econophysics have been integrated into a new group of Romanian teachers and researchers in economics and physics, gathered around workshops published as 'Exploratory Domains of Econophysics News' (EDEN), whose chairmen are the coordinator and one of the authors of this book, Gheorghe Săvoiu, and, Ion Iorga Simăn, respectively, of the University in Piteşti. The workshop initiated a journey into the fascinating world of trans- and multidisciplinary econophysics, which has gone on uninterruptedly up to the present and is available on the site http://ccma.upit.ro/.

This group, which included Mircea Gligor, Radu Chişleag, as well as Constantin Andronache and Aretina-Magdalena David-Pearson and other Romanian scientists based abroad, gradually expanded and managed to publish a further essential book entitled *Exploratory Domains of Econophysics* News (EDEN I and II), in 2009 (Săvoiu et al., 2009). Later, from 2011, it published the first online econophysics journal in Romania (http://www.esmsj.upit.ro/), now producing its fourth issue.

Worldwide, econophysics appears as either controversial (Cho, 2009) or extended across domain frontiers, toward philosophy (Rickles, 2007), or even admitting its affiliation with contemporary economics (Walstad, 2010), when it proves able to find much better solutions and much more relevant models, centered on the experimentalism and pragmatism of physics (Schinckus, 2010).

Debating the role and the potential of econophysics for Romanian scientific research is now not only an opportunity but a necessity for the normal evolution of both teaching and research in physics and economics in Romania, in universities and modern research centers.

REFERENCES

Andronache, C., Chişleag, R., Costea, C., David-Pearson, A-M., Ecker − Lala, W., Gligor, M., et al., 2009. Exploratory Domains of Econophysics News EDEN I &II. University Publishing House, Bucharest.

Ball, P., 2004. Critical Mass: How One Thing Leads to Another. Farrar, Straus and Giroux, New York, NY.

Becker, G., 1976. The economic Approach to Human Behavior. University of Chicago Press, Chicago, IL.

Blaug, M., 1992. The Methodology of economics: Or How economists Explain. Cambridge University Press, Cambridge.

Bouchaud, J.-P., Cizeau, P., Leloux, L., Potters, M., 1999. Mutual attractions: physics and finance. Phys. World 12 (1), 25−29.

Burda, Z., Jurkiewicz, J., Nowak, M.A., 2003. Is econophysics a solid science? Acta Phys. Pol. B 34 (1), 87−131.

Cho, A., 2009. Econophysics: still controversial after all these years. Science 325 (5939), 408.

Chakrabarti, K., 2005. Econophys-Kolkata: a short story. In: Chatterjee, A., Yarlagadda, S., Chakrabarti, B. (Eds.), Econophysics of Wealth Distributions. Springer, Milan, pp. 225−228.

Chatterjee, A., Yarlagadda, S., Chakrabarti, B., 2005. Econophysics of Wealth Distributions. Springer, Milan.

Di Matteo, T., Aste, T., 2007. No worries: trends in Econophysics. Eur. Phys. J. B: Condens. Matter Complex Syst. 55 (2), 121−122.

Fischer, J., Scholes, M., 1973. The pricing of options and corporate liabilities. J. Polit. Econ. 81 (3), 637−654.

Gallegati, M., Keen, S., Lux, T., Ormerod, P., 2006. Worrying trends in econophysics. Phys. A: Stat. Mech. Appl. 370 (1), 1−6.

Georgescu-Roegen, N., 1971. The Entropy Law and economic Process. Harvard University Press, Cambridge, MA.

Gligor, M., Ignat, M., 2001. Econophysics: a new field for Statistical Physics? Interdiscip. Sci. Rev. 26 (3), 183−190.

Gopikrishnan, P., Stanley, H.E., 2003. Econophysics two-phase behaviour of financial markets. Nature 421 (6919), 130−131.

Gordon, R.R., 2000. Reconciling Econophysics with macroeconomic theory. Phys. A: Stat. Mech. Appl. 282 (1), 325−335.

Guala, F., 2005. The Methodology of Experimental economics. Cambridge University Press, Cambridge, MA.

Jimenez, E., Moya, D., 2005. Econophysics: from game theory and information. Theory to quantum mechanics. Phys. A: Stat. Mech. Appl. 348, 505−543.

Kim, G.W., Markowitz, H.M., 1989. Investment rules, margin, and market volatility. Portfolio Manage. 16 (Fall), 45.

Krugman, P., Venables, A., 1990. Integration and the competitiveness of peripheral industry. In: Bliss, C., Braga de Macedo, J. (Eds.), Unity with Diversity in the European Community. Cambridge University Press, Cambridge, MA, pp. 56−75.

Lillo, F., Farmer, J.D., Mantegna, R.N., 2003. Econophysics: master curve for price-impact function. Nature 421 (6919), 129−130.

Mantegna, K., 1991. Levy walks and enhanced diffusion in milan stock exchange. Physica A 179, 39, 232–242.

Mantegna, R., Stanley, H., 2000. An Introduction to econophysics: Correlations and Complexities in Finance. Cambridge University Press, Cambridge, MA, pp. viii–ix.

Malkiel, B., 2001. An introduction to Econophysics: correlations and complexity to finance. J. Econ. Lit. 39, 142–143.

Mantegna, R.N., Stanley, H.E., 1997. Econophysics: scaling and its breakdown. J. Stat. Phys. 89 (1–2), 469–479.

Mirowski, P., 1989. More Heat than Light: Economics as Social physics. physics as Nature's economics. Cambridge University Press, Cambridge, MA.

Plerou, V., Gopikrishnan, P., Rosenow, B., Amaral, L.A.N., Stanley, E.H., 2000. Econophysics: financial time series from a Statistical Physics point of view. Phys. A: Stat. Mech. Appl. 279 (1), 443–456.

Reiss, J., 2007. Error in economics. Routledge, London.

Rickles, D., 2007. Econophysics for philosophers, studies in history and philosophy of science. Part B: Stud. History Philos. Modern Phys. 38 (4), 948–978.

Rosenberg, A., 1992. Economics—mathematical Politics or Science of Diminishing Returns. University of Chicago Press, Chicago, IL.

Schinckus, C., 2010. Is Econophysics a new discipline? The neopositivist argument. Phys. A: Stat. Mech. Appl. 389 (18), 3814–3821.

Stigler, G.J., 1964. Public regulation of the securities market. J. Bus. 37, 117–118.

Stanley, H.E., Afanasyev, V., Amaral, L.A.N., Buldyrev, S.V., Goldberger, A.L., Havlin, S., 1996. Anomalous fluctuations in the dynamics of complex systems: from DNA and physiology to Econophysics Phys. A: Stat. Mech. Appl. 224 (1–2), 302–321.

Stanley, H.E., Amaral, L.A.N., Canning, D., Gopikrishnan, P., Lee, Y., Liu, Y., 1999. Econophysics: can physicists contribute to the science of economics? Phys. A: Stat. Mech. Appl. 269 (1), 156–169.

Stauffer, D., 2000. Econophysics—a new area for computational Statistical Physics? Int. J. Mod. Phys. C 11 (6), 1081–1087.

Takayasu, H., Miara, H., Hirabuyashi, T., Hamada, K., 1992. Statistical Properties of Deterministic threshold Elements – The Case of Market Prices,. Physica A 184, 127–134.

Yakovenko, V.M., 2009. Econophysics, statistical mechanics approach to. In: Meyers, R.A. (Ed.), Encyclopedia of Complexity and System Science. Springer, Berlin.

Walstad, A., 2010. Comment on "econophysics and economics: sister disciplines?" by Christophe Schinckus. Am. J. Phys. 78 (8), 789.

Wang et al., 2004. Physics of econophysics. Working Paper of Beijing Normal University, no. 1025. http://arxiv.org/pdf/condmat/0401025.pdf. Last accessed: The 2nd of October, 2012.

FURTHER READING

McCauley, J.L., 2006. Response to worrying trends in Econophysics. Phys. A: Stat. Mech. Appl. 371 (2), 601–609.

McCauley, J.L., 2004. Dynamics of Markets: econophysics and Finance. Cambridge University Press, Cambridge, MA.

CHAPTER 2

Multidisciplinary Modeling Knowledge and Unidisciplinary Isolation

Gheorghe Săvoiu and Ion Iorga Simăn
University of Piteşti, Faculty of Economics and Faculty of Sciences, Romania

2.1 INTRODUCTION
2.2 MULTIDISCIPLINARY MODELING
2.3 THE PRINCIPLES OF MULTIDISCIPLINARY MODELING
2.4 A FINAL REMARK
REFERENCES

2.1 INTRODUCTION

The term science, in the sense of knowledge, is derived from Latin *scientia* and can be encapsulated as a systematic ensemble of knowledge connected with nature, society, and thinking. Scientics or scientology currently represents the science of science, an investigation into the way in which the study of nature through observation and reasoning has evolved all through several millennia of human activity. Logic is, in its capacity, a mode of thinking about thinking, achieving unanimous recognition as the first science in the history of sciences. Mathematics has come as a result of the studies on quantities and hierarchies, turned into theorems by means of logical derivation, to be called a science of quasi-general usefulness, yet, without physics and its necessary limits and aspect of finiteness, introduced into mathematical reasoning, the results of scientific knowledge would rather be axiomatic systems of infiniteness. Through methodically measuring the manner in which the characteristics of populations vary, statistics rounds up logics, mathematics, and physics, while emphasizing the importance of observation and reasoning, in much the same way as physics does, by means of experiment and simulation, in its perpetual attempt to grasp reality. And so, the broad spectrum of natural science is reached, where science describes a systematic study, or the knowledge acquired subsequent to that study conducted on nature, starting from human nature (anatomy, sociology, etc.) up to animal, and even inanimate, nature (biology, geology, etc.). The expansion of contemporary science has multiplied their number to over 1,000 independent sciences, especially within borderline areas (e.g., econophysics, situated at the border between physics and economics, sociophysics—at the border between physics and sociology, etc.). Science emerges when at least three elements are joined together: a distinctive theory, a segment of reality as a specific object, and a model interposed between theoretical investigation and its object of study. Sciences have their own characteristic models and laws, acquired mainly thanks to their inclination for measuring their object of study.

2.2 MULTIDISCIPLINARY MODELING

Logical representation is the solution preceding the model. From the tetragrams of the ancient Chinese culture to abstractization through geometrical figures in ancient Greece, from the first music sol-fa systems to the everyday languages used by calculus programs, all these types of presentation laying special emphasis on the logical element of a visual nature have been, and are still, simplified alternatives to modeling.

At present, modeling is simultaneously recognized as a method, and a component part of that triad making up sciences (Figure 2.1).

In a relevant way, the model and modeling have been situated, through their initial practical uses, closer to geometry than any other scientific domain. From the very beginning, the fifth postulate, conceived by Euclid in his *Elements*, was regarded with suspicion by his Hellenic contemporaries. In an unpublished paper, which was only shared with friends, Gauss underlined the fact that, starting from statements that contradicted the permanently "inculpated" postulate, one could develop a compatible geometry. Lobacevski, Bolyai, and Riemann each created original non-Euclidean geometries compatible with Euclid's geometry, where no pair of parallels existed any longer. The appearance of the term as such is linked to the year 1868, when the mathematician Eugenio Beltrami managed to construe an early Euclidian model for non-Euclidian geometry. For the first time, he

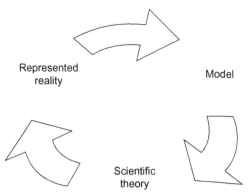

Figure 2.1 The method of modeling.

was turning the model and modeling into a concept, studying, by their agency, a domain, a phenomenon, an object inaccessible to direct research. The geometry-inspired model became "a coagulant factor" for scientific thinking, a continuous process of pondering, represented, symbolized, and conveyed, no less than the tetragrams were to Gottfried von Leibniz the inductive solution to the mechanic system of his own calculating mechanical device. At a higher level of elaboration, models are scientific representations or representations of scientific theories. Paraphrasing Parmenides, the model that can be thought and the one for which the thought exists are one and the same. Theoretical science, a permanent source of experimental suggestions, becomes at once experimenting and foreseeing, and along these lines the basic conditions of multidimensional modeling can be synthesized as follows:

- The first condition for a model is its direct relationship with thinking ("a bird is a machine functioning in accordance with the laws of mathematics, an instrument that man can reproduce with all its motions"—to quote Leonardo da Vinci, in *Macchine per volare*).
- A second condition is the identification of the essential aspects and formulating questions in a correct manner.
- The profoundness, the intensity, and the depth represent the third condition of representation through models (the oscillation between analogy and the convention symbol).
- The efficiency of the transposition or the translation of the theory into the reality of the world under study seems to be another condition, the superior models becoming themselves objects of research

and remodeling. The researchers who confine themselves to a sole model burden themselves with an even greater sin, as Georgescu-Roegen (1971) remarked: "the sin of complete ignorance of the qualitative factors that cause endogeneous variability" (Georgescu-Roegen, 1971).

In keeping with the reasoning of modeling, as maximum fidelity translation or transposition, any theory corresponds to a model, and any model, when validated through the agency of reality, will correspond to reality. However, the closer the model will draw to the point of intersection of several sciences, the more correct the transposition/translation. Even the exclusive answer to the question "what is a model" constitutes a difficult undertaking and needs many-sided approaches. Below are some illustrative variants:

- In the optima of physics, a model is a calculating instrument, with the help of which one can determine the answer to any question concerning the physical behavior of the system in question, or else a precise pattern of a certain segment of the physical reality (two examples, which are today as well known as to become banal, are the modeling of the inertial reference system and the atomic model).
- In the vision specific to chemistry, the model becomes a structural concept that attempts to explain the properties found experimentally or a support in deductively passing from the general to the specific, a knowing instrument that forecasts facts and "indicates the numbers" (as in the memorable example of Mendeleev's table of elements or the periodicity of chemical elements).
- In the approach of biology (genetics), a model is considered a natural modality—reproduced experimentally—of genetically differentiating the populations (the model of DNA being, in this respect, a commonly cited example and a relevant case in point).
- In the perspective of mathematics, the model is superposed to a certain type of measuring methods, specific to mathematical research, with a view to explain, in an objective manner, the "manner in which the microcomponents and their mutual interactions either interpreted individually or grouped in subsystems, generate and explain the whole of the system" (Octav Onicescu and the model of informational energy), or a "definition and noncontradictory description of a number of processes and phenomena," of the theses,

postulates, and axioms, as well as their logical—mathematical correspondence.

- From a logical point of view, within the structure of the model, the causes equalize the effects (Dumitriu, 1998).
- From a behaviorist standpoint, the model presupposes a number of participants gathered in a formal way, who "maximize their utility by starting from a stable set of preferences and accumulate an optimal amount of information in a variety of markets" (Becker, 1990).
- Along the lines of the semantic, linguistic, and explanatory dominant, the model is a theoretical or material system by means of which one can study, indirectly, the properties and transformations of a different, more complex system, where the first system exhibits an analogy (according to the explanatory dictionary).
- In its statistical connotation, the phase-directed sense of the concept of model is that of a link in an integrated process of knowing and is made up of a hypothesis, a schematic representation of a process (phenomenon), the statistical testing, and the resuming of the process in a general theory.
- In keeping with modern sciences, the multidisciplinary model becomes the optimum instrument for solving a number of complex general problems, and modeling turns into a series of means meant to disclose the real nature of the problems, where the isolated vision does not allow one to formulate characteristic laws.
- The statistical or physical—mathematical type of modeling is a mathematical transcription of a number of simplified hypotheses about the state or evolution of a social—economic phenomenon, or physical system under the factorial influence of variables that are physical or can be assimilated to the physical ones.

The multidisciplinary model turns to account the language and methods of mathematics, testing, and statistical decision, the pattern of physics in assessing (quantum, thermodynamic, acoustic, etc.) reality, as well as the real variables of the segment subject to research (money flow in the economy, human behavior in sociology, etc.)

How can one manage to practically construct a model? The starting point is direct experience or unmediated contact with reality. In order that a theory could be turned an experiment, or into an "organized contact with reality," a theory is formulated, which is subsequently represented by a material, intuitive or symbolic model, as a filtered reflection of reality. Louis Pasteur would elegantly underline the primacy of the theory, through the agency of the well-known formula: "luck favors only the well-prepared mind." Tiberiu Schatteles used to synthesize the likeness between theory and modeling through the phrase "the dogmatics of isolation." In order to illustrate a phenomenon, the theory isolates it from the contingent, very much as the experiment is underlain by a type of material (i.e., laboratory) isolation. Studying a phenomenon in isolation also presupposes defining the framework of the isolation through postulates or axioms as "something that goes without saying." Modeling, as a complex iterative process, oscillates between simplified variants like the "triad" (formulating a hypothesis, collecting the experimental material, and verifying the hypothesis) and excessively detailed variants (formulation of the initial model followed by the forming of repartition classes, gathering the experimental material or the data, choosing a particular repartition, checking the degree of concordance of the repartition chosen with the real situation, and formulating the hypotheses that explain the random mechanisms that have generated the data). The typological diversity of the models results from the great number of the scientific theories that they reproduce. Seen from the angle of the aim they were created for, the models fall into two major types: the category of the rational or theoretical models, and the category of the operational models or prediction (decision-making) models.

Through comparison with the time variable, modeling is static or dynamic. A major classification of modeling according to the typology of the explanatory variables reveals the deterministic type of modeling in the past (evolution of phenomena, determined solely by the mechanical, or simply causal variables) and modern probabilistic modeling (which contains perturbing variables, in keeping with the probable effect of some uncontrolled factors and unspecified variables). Exploratory Domains of Econophysics News I − EDEN I led all the participants to the idea that modeling can be unidisciplinary, but it will remain isolated in the past, as well as up-to-date, i.e., covering reality, and, implicitly, multidisciplinary. For a succinct description of multidisciplinary modeling, a few clarifications are in order, relating to the various stages, its architecture and paradoxes. The concrete stages of modern modeling are the following:

1. The structural defining of the system (isolating the phenomenon, formulating the questions, identifying the major interest variables).

2. The preliminary formulation (sets of hypotheses and conclusions concerning the relationships between the variables), collecting the empirical (relevant) data.
3. The estimation of the parameters and of the functional forms.
4. The preliminary (gross) testing.
5. The additional testing (based on the new data).
6. The decision—accepting or rejecting (in conditions of predictions conforming or failing to conform to the available empirical evidence).

The architecture of multidisciplinary modeling capitalizes on:

a. Minimal simplification through hypotheses, formulated for the first time by William Ockham (the economic architecture, or law of parsimony), or the existence of a minimal number of propositions not connected mutually and undemonstrated propositions ("out of two interpretations of a phenomenon, the interpretation having fewer suppositions or simplifying hypotheses is preferred)".
b. The simple alternative (the highly intricate models failed to lead to categorically better results, as against the simple extrapolation formulas—T.C. Koopmans).
c. The value certified through the dialectical reasoning (a model facilitates the discussion, clarifies the results and limits the reasoning errors).
d. The cultural component (if humans' economic and social actions were independent of their cultural inclinations, the enormous variability of the economic, and social configuration in point of time and place could by no means be accounted for).
e. Shifting from one-discipline to multidisciplinarity, through successive models (improvement through imitation, analogy, and passing from one type to another).

Multidisciplinary modeling is a process having a paradoxical content. The paradox of the infinity of the multivariable system is revealed by the infinite number of factors, which cannot be classified in a direct manner, in proportion to the particular model construed out of a finite number of essential factors.

The paradox of the "relative reduction of one system to the next" proceeds from relative reductibility, centered on the translatability of the languages concerning various fields of reality and manifests itself as an antithesis between the functional and the substantial.

The paradox of the "unique community" can be translated through the antinomy holding between the correlation of the action of several models and the building up of a unique model for a given problem.

The paradox of the "double idealization" concerns the phases of simulation, and respectively, of the assignation and interpretation of information within the model. Multiplied information is not lost from the model, as, fragmented, it is nothing but information.

The "double idealization" consists in treating information as "signification" of information in the model of attribution, whereas in the interpretation model it is treated equally as signification and as sense.

Synthetically, the relationship between completeness and precision/accuracy generates specific models (Table 2.1).

The algorithm of the model has three characteristic features: determinism in point of performance, succession in point of operation, and universality in so far as the spatial, temporal, and structural entries and limitations are concerned. Modeling exhibits three main ways of analysis:

1. Using the equilibrium equations between the factors (Leontief's input–output balance or Leontief's model was subsequently generalized in three distinct variants, i.e., the deterministic, the random, and the information ones, to the fuzzy ones, and to those of quantum physics, thermodynamics, etc.).
2. Identification of the extreme values as the model of the "catastrophes," or of R. Thom's "critical points"—which is the frequently cited example in point.

Table 2.1 Degree of Completeness and Precision of the Data Generating the Typology of the Model (Săvoiu, 2001)

Degree of Completeness of the Data	Degree of Precision of the Data	Typology of the Model
Maximum	Maximum	Deterministic
Relatively low	Relatively high	Probabilistic
Relatively high	Relatively low	Fuzzy
Relatively low	Relatively low	Intuitive
Minimum	Minimum	Indeterministic

3. Construction (simulation) of conflict situations through strategic games with incomplete information (i.e., concurrent situations) or complete information (i.e., open situations).

The uncertainty of decision making is paramount, all the way from Wald's (prudent or pessimistic) model, characterized by choosing the maximum profit variant, or the minimal loss cost wise, in the most unfavorable situation, to Laplace, which selects the higher average-profit variant, or the lower average-loss, in the hypothesis that the states have the same occurrence probability, to Savage, where an option is made for the lowest possible regret (i.e., the usefulness lost as a result of selecting a different variant than the optimal one, in conditions of complete information), and to Hurwicz, whose coefficient of optimism reenters, through its real-value interval, the vast realm of the probabilities, namely [0,1]. To illustrate the above, multidisciplinary modeling maximizes the capacity of reducing the degree of *imprecision/inaccuracy* and of assessing that *imprecision/inaccuracy* through statistical testing and testing in terms of probability theory, whereas even mathematical modeling approximates, while failing to express reality exactly as it is, because reality is not "exact/precise," but subject to the stochastic laws or to the action of the law of great numbers.

To express in a multidisciplinary manner how inaccurate a model is, is more important than modeling in a unidisciplinary, hence sophisticated, isolated manner, lacking the power of specificity. The perspectives of the field of model construction astonish through the rigor of a new concept, namely that of the system of multidisciplinary models, which presupposes the following principles:

- The human decision has the fundamental role in its functioning.
- The construction is a logical succession and also a process of arrangement in time, in keeping with the principle of economy or the law of parsimony.
- The separation and combination of the individual models occurs in procedure-based chains.
- The system stays open, thus facilitating the adding or the deletion of restrictions and variables.
- The physical–mathematical structure is independent of the manner of utilization.
- The architecture is modular, hierarhical, and dynamic.

- The information-based and logical connections are, in turn, part of cooperative, hierarchical, mixed models.
- Although including different types of models, the database is unique.

In the natural harmony of the multidisciplinary approach to modeling, the contribution scored by discovering of an original model is to be considered much higher than knowing a new phenomenon or process (Pecican, 2003).

The limits of unidisciplinary modeling are obvious:

- No unidisciplinary model can consistently and substantially incorporate the residual variables and areas (which can occasionally be quite considerable in point of proportions and significations).
- Both human behavior and other random variables like the climate, radical political evolutions such as revolutions, as soon as they are modeled, bestow an increased amount of uncertainty to the respective model.
- The model has evolved in a credible manner along the coordinates of the chronological series, and less so, however, along those of the territorial series, of the associated/correlated series, in the specific situations of value optimizations, or concerning verisimilar, attainable targets developing programs.

To conclude, a model can be said to represent an image of a specially selected part of reality, with the aid of which answers can be given to various questions, or problems belonging to an assortment of fields in the area of scientific knowledge can be solved, with a certain degree of realism and a certain limit of error. The main disadvantage of the unidisciplinary model, if one resorts to the example provided by the very econometric one, is revealed by the lack of accuracy of their prediction, by the representatives of the neoclassical Austrian school of economics, Ludwig von Mises (1966) and Friedrich von Hayek (1989). The sad balance of the predictions made by the econometric models over the past few years, for all the modern calculation equipment added to the sophisticated unidisciplinary models, is nothing but an additional confirmation (von Mises, 1966; Hayek, 1989).

2.3 THE PRINCIPLES OF MULTIDISCIPLINARY MODELING

Certainly, any research admits, like Socrates, the impossibility of absolute modeling knowledge. The famous "ignoramus et ignorabis" (*we don't know and we shall never know*) belonging to Emile du Bois-Reymond is a statement containing a relative truth. Yet, as truth always thrives on liberty, let us enumerate a few of the principles of multidisciplinary modeling, as they appeared in the papers presented at EDEN I and in the researchers' dialogues:

1. There is a harmony of modeling disagreements, a concord of discordances, a diversity revealing of the unity of the model.
2. The developmental cycle is the axis of the cyclical development in the model.
3. The motion through an apparent state of rest and the state of rest of the motion are the realities of all the cases of modeling. As a paraphrase to one of Schlozer's dictums, science remains history at rest, very much as history becomes science in motion. And the scientific modeling knowledge of the multidisciplinary type seems to be defining, in an increasingly clear manner, a solution to the famous "ignoramus et ignorabis."
4. The identification of the leap, or the unpredictable transformation, in the sense of the paradox of the arrow that slays Achilles, or of the tortoise which overtakes the hare, represents the spirit of modeling.
5. Communication, as an aim of getting out of information isolation, constitutes the message of the types of modeling.
6. The relativity of the global interdependencies and of the local ones derives from the logic of the systems modeled, namely when the sum of the parts is greater than the whole.
7. The infinite, as part of the finite, and the finite, as part of the infinite, describe the structures of the model.
8. The finality of the inductive through deduction and the validation of the deductive through induction bound the reasoning of those who do the modeling.
9. Knowledge is the limit to the ignorance of the act of modeling, no less than ignorance eventually becomes the result of knowledge.
10. The rebirth of theory through experiment brings about the demise of experiment in the theory of modeling.

11. The faith in critical science becomes similar to the neutrality of ignorance in the acts of modeling.
12. Coherent superposition brings together the amplitudes as limits, while incoherent superposition unites only the intensities through modeling.
13. Finding nuances is a solution of probabilistic thought and based on the possibilities of modeling.
14. Convergence through divergence contributes to the emergence of the models.
15. The incompleteness of completeness adds to the completeness of incompleteness in modeling.
16. The compensation of the reactions confers equilibrium to imbalance.
17. The duality of the acts of modeling is a reflex of the equivalence causes–effects.
18. A fixed multidisciplinary modeling method is not really a method.

2.4 A FINAL REMARK

To conclude, multidisciplinarity is unifying, while unidisciplinarity isolates. Thence, the culture of multidisciplinary modeling remains a practical issue, not certainly in as far as that culture is regarded only as a product of life, but life (reality) having become, in that sense, a consequence or an imprint of culture.

REFERENCES

Becker, S.G., 1990. The Economic Approach to Human Behavior. 2nd Ed. University of Chicago Press, Chicago, IL.

Dumitriu, A., 1998. The History of Logic. Technical Publishing House, Bucharest.

Georgescu-Roegen, N., 1971. The Entropy Law and Economic Process. Harvard University Press, Cambridge, MA.

Hayek, F., 1989. The Collected Works of F.A. Hayek. University of Chicago Press, Chicago, IL.

Pecican, E.Ş., 2003. Econometrics for …Economists. Economical Publishing House, Bucharest.

Săvoiu, G., 2001. The Price Universe and the Interpret Index. Economical Independence Publishing House, Piteşti.

von Mises, L., 1966. Human Action: A Treatise on Economics, 3rd rev. ed. Henry Regnery and Co., Chicago, IL.

PART II

Economics and Finance

CHAPTER 3

The Efficiency of Capital Markets: Hypothesis or Approximation

Gheorghe Săvoiu[1] and Constantin Andronache[2]
[1]University of Pitești, Faculty of Economics, Romania; [2]Boston College, Chestnut Hill, MA, USA

3.1 INTRODUCTION: DEFINING MARKET EFFICIENCY

The efficient market hypothesis (EMH) is getting new consideration in recent studies of financial markets in part due to the need for a baseline price model and in part due to the practical aspects of trading and investing.

Informational efficiency: Market efficiency has varying degrees: strong, semistrong, and weak. Stock prices in a perfectly efficient market reflect all available information. These differing levels, however, suggest that the responsiveness of stock prices to relevant information may vary. The EMH states that a market cannot be outperformed because all available information is already built into all stock prices. Practitioners and scholars alike have a wide range of viewpoints as to how efficient the market actually is.

Our study is concerned only with aspects of the informational efficiency of the capital markets. Some of the significant contributions to the "market efficiency" concept, asserting that EMH is true, are summarized here:

- Bachelier (1900): In his PhD thesis:*"past, present and even discounted future events are reflected in market price, but often show no apparent relation to price changes."*
- Samuelson (1973): In the article, "Proof that properly anticipated prices vibrate randomly," stated: *"...competitive prices must display price changes...that perform a random walk with no predictable bias."* Therefore, price changes must not be predictable if they are properly anticipated.
- Fama et al. (1969): First definition of *"efficient market is a market which adjust rapidly to new information."*
- Fama (1970): *"A market in which prices always 'fully reflect' available information is called 'efficient'."*
- Rubinstein (1975) and Latham (1985): They have extended the definition of market efficiency. The market is said to be efficient with regard to an informational event if the information causes no portfolio changes.
- Jensen (1978): He states that prices reflect information up to the point where the marginal benefits of acting on the information (the expected profits to be made) do not exceed the marginal costs of collecting it.
- Malkiel (1992): He offered the following definition: *"A capital market is said to be efficient if it fully and correctly reflects all relevant information in determining security prices. Formally, the market is said to be efficient with respect to some information set...if security price would be unaffected by revealing that information to all participants. Moreover, efficiency with respect to an informational set...implies that it is impossible to make economic profits by trading on the basis of (that informational set)."*

3.1.1 Criticism of the "Market Efficiency" Concept

Some of the contributions of EMH criticism and exposures of its limitations, concluding that EMH in strict sense is not possible in real financial markets, are summarized here:

- Grossman (1976) and Grossman and Stiglitz (1980) argue that perfect informational efficient markets are an *impossibility*, for if markets are perfectly efficient, the return to gathering information is nil, in which case there would be little reason to trade and markets would eventually collapse.

- Campbell et al. (1997) are in favor of the notion of relative efficiency—the efficiency of one market measured against another.
- Lo and MacKinlay (1999) say: "...*the Efficient Markets Hypothesis, by itself, is not a well-defined and empirically refutable hypothesis. To make it operational, one must specify additional structure, e.g., investors' preferences, information structure, business conditions, etc. But then a test of the Efficient Markets Hypothesis becomes a test of several auxiliary hypotheses as well, and a rejection of such a joint hypothesis tells us little about which aspect of the joint hypothesis is inconsistent with the data.*"
- Fama's revision (1991): Efficiency *per se* is not testable. It must be tested jointly with some model of equilibrium. When we find anomalous evidence on behavior of returns, the way it should be split between market inefficiency or a bad model of market equilibrium is ambiguous.
- Zhang (1999): "*Empirical evidence suggests that even the most competitive markets are not strictly efficient. Price histories can be used to predict near future returns with a probability better than random chance. Many markets can be considered as favorable games, in the sense that there is a small probabilistic edge that smart speculators can exploit.*"
- Blakey (2006): "*If academics had introduced the efficient market approximation, rather than the efficient market hypothesis, years of pointless debate and a huge schism between academics and practitioners would both have been avoided.*"
- Beechey et al. (2000): "*The efficient market hypothesis is almost certainly the right place to start when thinking about asset price formation. The evidence suggests, however, that it cannot explain some important and worrying features of asset market behaviour. Most importantly for the wider goal of efficient resource allocation, financial market prices appear at times to be subject to substantial misalignments, which can persist for extended periods of time.*"

3.1.2 Types of Market Informational Efficiency

Weak-form efficiency: the information set includes only the history of prices or returns themselves. A capital market is said to satisfy weak-form efficiency if it fully incorporate the information in past stock prices.

Semistrong form efficiency: the information set includes all information known to all market participants (*publicly available information*). A market is semistrong efficient if prices reflect all publicly available information.

Strong form efficiency: the information set includes all information known to any market participant (*private information*). This form says that anything that is pertinent to the value of the stock and that is known to at least one investor is in fact fully incorporated into the stock value.

3.2 TEST OF MARKET EFFICIENCY

3.2.1 Weak from Tests

How well do past returns predict future returns? The main assumption of the EMH is that there is no pattern in the time series of returns, such that the returns should be approximated by a random walk. A consequence is that the autocorrelation of the returns time series should be negligible. In literature, there are several ways to test price behavior, such as: (a) the fair-game model, (b) the martingale model, and (c) the random walk model. In our approach, we will test for random walk, i.e., we will concern ourselves with testing to see if the daily returns from a major market index data can be approximated with a random walk, or perhaps there are significant departures.

3.2.2 The Random Walk Model

The simplest form version is the independently and identically distributed (IID) increments case, in which the dynamics of prices are given by a random walk. The independence of increments implies not only that price increments are uncorrelated, but that any nonlinear functions of the increments are uncorrelated. This model is known as Random Walk 1. The assumption of identically distributed increments is not plausible for financial asset prices over long time spans, due to changes in the economic, social, technological, institutional, and regulatory environment in which stock prices are determined.

By relaxing the assumptions of the random walk to include processes with independent but not identically distributed (INID) returns, we have the model Random Walk 2. This model still retains the most interesting property of Random Walk 1: Any arbitrary transformation of future price increments is not predictable using any arbitrary transformation of past price increments. By relaxing the independence assumption of Random Walk 2 to include processes with dependent but uncorrelated increments, we obtain the weakest form of random walk, named Random Walk 3 (in our study we will test for this form of random walk).

3.3 DATA AND PROCEDURE

Charles Dow and Edward Jones gave their names to the most famous and followed financial index in the world, after they formed Dow Jones & Company. The company was delivering financial information to those who needed it (Rosenberg, 1982). The first news sheet of Dow Jones & Company was printed in 1883 and was the forerunner of *The Wall Street Journal*. Charles Dow's observations are considered some of the most important earlier writings relating to technical analysis. There are many arguments of why technical analysis works when applied correctly to trading any financial market, but the most important sound like a question: "If it were not possible to make money trading because the markets are inherently random, then why do so many traders make money?" Technical analysis is the science of human behavior. If you are in tune with the market sentiment, then you can trade this knowledge effectively. Technical analysts try to identify and exploit possible market inefficiencies, at least in the short term.

The Dow–Jones industrial average (DJIA) represents a barometer of a major capital market (30 largest with high-consistent performance corporations of US economy). To calculate the DJIA, the sum of the prices of all 30 stocks is divided by a divisor, the DJIA divisor. The initial divisor was the number of component companies, so that the DJIA was at first a simple arithmetic average; the present divisor, after many adjustments, is less than one (meaning the index is actually larger than the sum of the prices of the components):

$$\text{DJIA} = \frac{\sum p}{d} \tag{3.1}$$

where p denotes the prices of the component stocks and d is the Dow divisor.

In the case of splits or changes in the list of the companies composing the index in order to avoid discontinuity in the index, the Dow divisor is updated so that the quotations right before and after the event coincide:

$$\text{DJIA} = \frac{\sum p_{\text{old}}}{d_{\text{old}}} = \frac{\sum p_{\text{new}}}{d_{\text{new}}} \tag{3.2}$$

The DJIA is only one index from a large indices' family, and it has chronicled more than 113 years of investing and has served as a marker through all of the major developments in modern history.

The index used in this study covers the time interval 1928–2008, daily closing values. The data employed consists of daily logarithm returns of the DJIA index. It represents a highly traded market, such that large amount of information is available to all economic agents. It represents a good "barometer" of financial markets since the DJIA index correlates quite well with other major market indices. DJIA provides a challenge for the test of EMH because it is expected that most information is available to most players.

Figure 3.1(a) shows the main features of the daily evolution of the DJIA from 1928 to 2008. In log scale, the index shows a consistent growth over decades, caused by population and wealth growth,

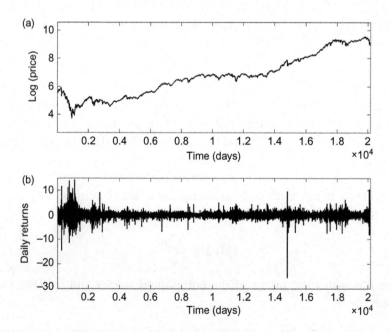

Figure 3.1 (a) The natural logarithm of the daily close value of DJIA, log[y(t)] vs. time and (b) daily logarithmic returns of the DJIA (expressed as percentage), defined as r = 100 × log(y$_{(t+1)}$/y$_{(t)}$). Time interval: 1928–2008.

reflected in capital market participation. In linear scale (not shown), the growth is exponential, with very rapid growth in the last three decades. In Figure 3.1(b), the daily logarithmic returns are shown. Daily logarithmic returns of the DJIA (percentage) are defined as:

$$r = 100 \times \log \frac{y_{(t+1)}}{y_{(t)}} \qquad (3.3)$$

Large variations are visible especially during major market corrections and crashes such as 1929, 1987, and 2008. Another noticeable feature is the clustering of variability (also linked to volatility or variance of returns).

Figure 3.2 shows the normal probability plot of daily logarithmic returns of DJIA, r, as defined above. The dotted line corresponds to a normal distribution, while the solid curve corresponds to calculated daily returns from DJIA data. The plot suggests the daily logarithmic returns of DJIA values are not normally distributed and have "fat tails."

Figure 3.3(a) shows the autocorrelation function (ACF) of daily returns of DJIA, r, versus lag (in days) and (b) show the partial autocorrelation function (PACF) of daily returns of DJIA, r, versus lag. Plots suggest that autocorrelation of daily return is marginally

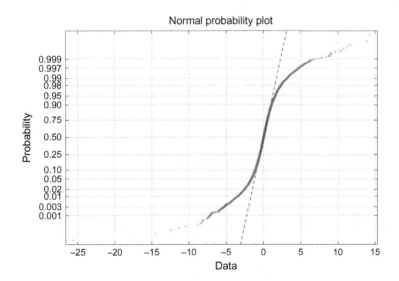

Figure 3.2 Normal probability plot of daily logarithmic returns of DJIA, r, as defined in text: dotted line corresponds to a normal distribution, while the solid curve corresponds to calculated daily returns from DJIA data.

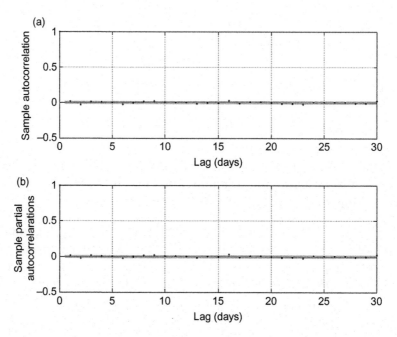

Figure 3.3 (a) ACF of daily returns of DJIA and (b) PACF of daily returns of DJIA. Plots suggest that autocorrelation of daily return is marginally significant (the horizontal lines represent the 95% CL).

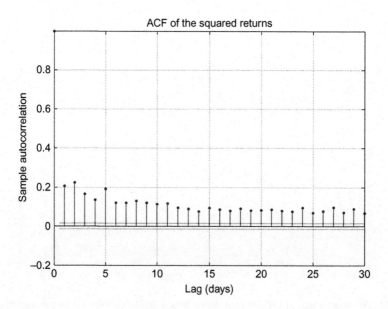

Figure 3.4 ACF of squared returns vs. lag (in days).

significant (the thick, horizontal lines represent the 95% CL). In Figure 3.4, the ACF of squared returns versus lag (in days) is displayed. It suggests statistically significant autocorrelation in daily volatility. While the ACF for daily returns indicates marginal significance, in the case of squared returns, the ACF indicates significant persistence of volatility. This forms the basis of GARCH (Generalized Auto-Regressive Conditional Heteroskedasticity) modeling of volatility of returns.

It suggests statistically significant autocorrelation in daily volatility.

Some interesting and significant differences that appear in the process of comparison of DJIA daily returns with random walk are shown in Figure 3.5: (a) daily DJIA price vs. time; (b) daily DJIA logarithmic returns vs. time; (c) daily logarithmic returns of random walk data vs. time; (d) test of normality for DJIA daily price; (e) test of normality for DJIA daily logarithmic returns, and (f) test of normality for daily

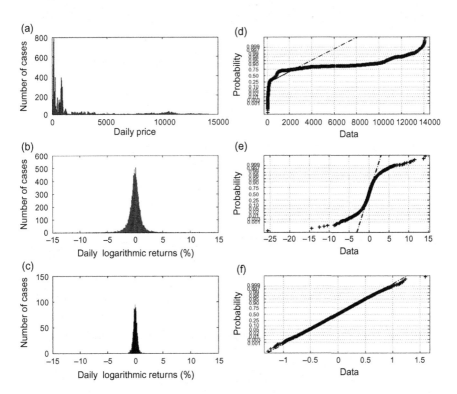

Figure 3.5 DJIA daily logarithmic returns and their specific probabilities: (a) 1928–2008 DJIA daily price; (b) 1928–2008 DJIA daily logarithmic returns; and (c) daily logarithmic returns of random walk.

logarithmic returns of random walk data. The difference between (e) and (f) suggests that real returns of DJIA are not fully described by a pure random walk.

3.4 VARIABILITY OF MARKET EFFICIENCY

We investigated the EMH validity during two contrasting time periods: (1) the interval 1928–1932 during the significant market decline of the Great Depression and (2) the interval 1980–2000, a period of positive trend, in which the market grew significantly. Table 3.1 gives that: (a) average return over all time intervals (1928–2008) is close to zero, while on the predominantly positive trend period (1980–2000) it is ~ 0.05. Thus, daily returns in uptrend produce profit on average. In contrast, during 1928–1932, during the predominant downtrend, the average daily return is -0.13. The standard deviation (STD) of returns varies also with the nature of the data, suggesting larger daily volatility during panic periods (1928–1932). The kurtosis is much larger than 3 (value corresponding to a normal distribution). Thus the daily returns have a distribution with a peak more pronounced than a normal distribution probability density function (pdf). The skewness suggests that the daily return distribution is not symmetric, with more likely large negative returns, except during 1928–1932, where the skewness is smaller. These considerations suggest that daily returns do not follow a random walk in strict sense, but the deviations are small. These small deviations from perfect efficient markets are somehow related to the technical analysis employed in financial markets and form the basis of trading schemes seeking superior performance.

Based on these preliminary results, we applied the same method for data divided by decade. While the division is somehow arbitrary, we note that a decade has sufficient data, some decades can be recognized as having a defined trend, while other have a mix of market behavior (Figures 3.6–3.14). The purpose here is to illustrate that EMH test against random walk process gives a significant variability by decade, reiterating that EMH is not satisfied in a strict sense (while it is still considered acceptable by some authors).

Table 3.1 Statistical Characteristics of Daily Logarithmic Returns of the DJIA Index				
Time Interval	Average	Standard Deviation	Kurtosis	Skewness
1928–2008	0.0175	1.1602	28.4400	−0.6147
1928–1932	−0.1300	2.6229	6.7497	0.0456
1980–2000	0.0521	1.0322	83.7348	−3.1886

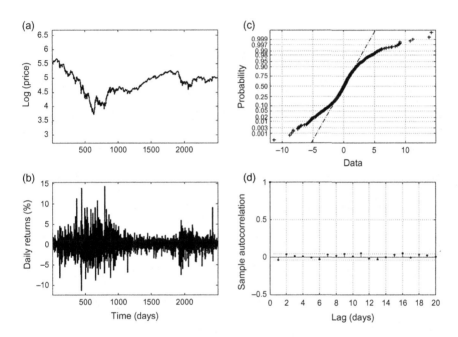

Figure 3.6 DJIA 1930–1939: (a) daily log(price) vs. time; (b) daily logarithmic returns (percentage) (r) vs. time; (c) test of normality of r; and (d) autocorrelation coefficients vs. lag (days) (the horizontal dotted lines are the 95% CL).

Significant variability is seen by decade in the average and STD; kurtosis is higher than 3, such that distribution has a higher peak than a normal distribution; skewness is mostly negative and the normality test shows that in most cases the distribution is likely not normal ($H = 1$). In few instances, $H = 0$ (distribution is likely normal) (Table 3.2).

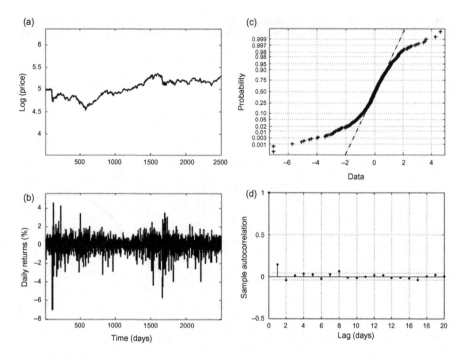

Figure 3.7 Same as Figure 3.6, for the time interval 1940–1949.

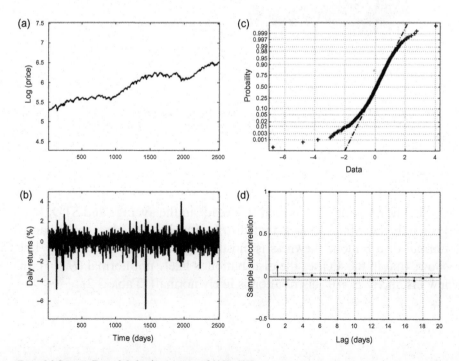

Figure 3.8 Same as Figure 3.6, for the time interval 1950–1959.

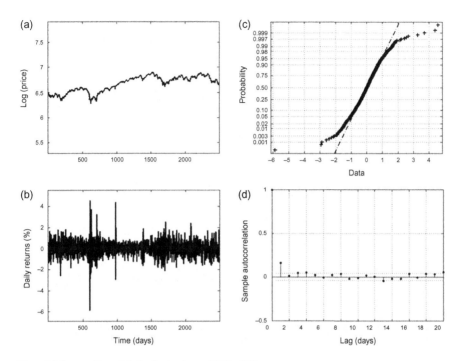

Figure 3.9 Same as Figure 3.6, for the time interval 1960–1969.

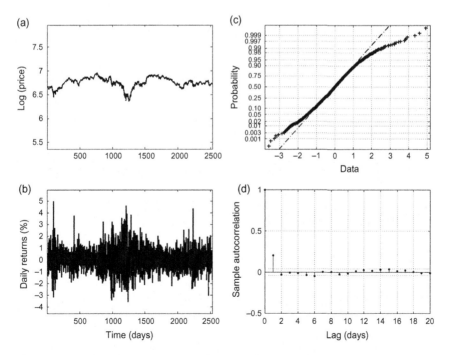

Figure 3.10 Same as Figure 3.6, for the time interval 1970–1979.

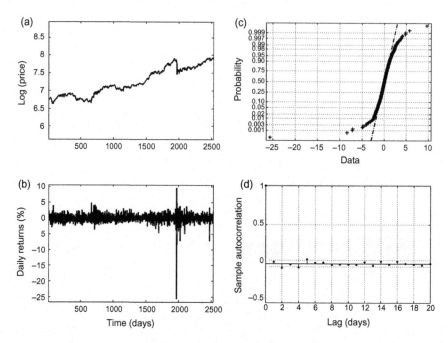

Figure 3.11 Same as Figure 3.6, for the time interval 1980–1989.

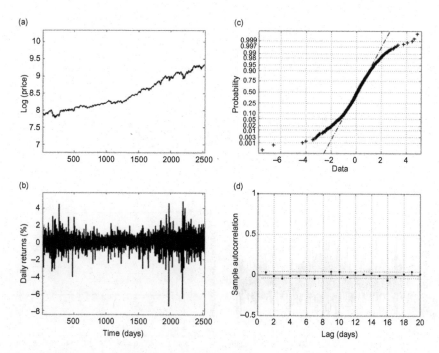

Figure 3.12 Same as Figure 3.6, for the time interval 1990–1999.

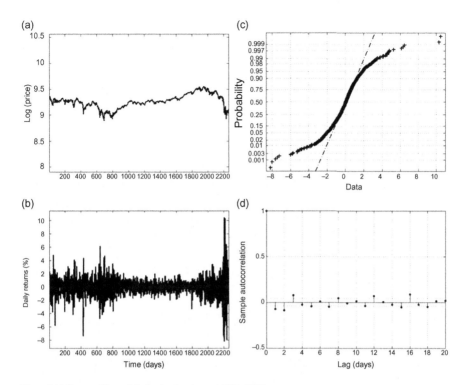

Figure 3.13 Same as Figure 3.6, for the time interval 2000–2008.

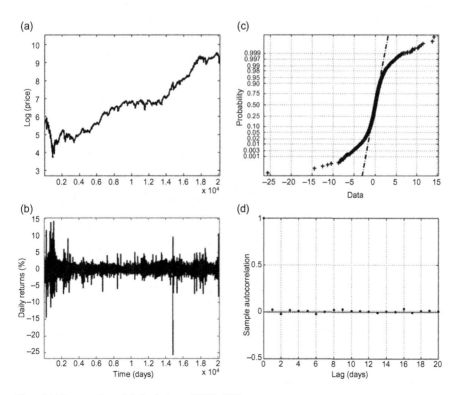

Figure 3.14 Same as Figure 3.6, for the interval 1928–2008.

Table 3.2 Statistics of the DJIA Daily Logarithmic Returns (%) (r) for the Detailed Intervals and for Whole Interval 1928–2008

Interval	Average	STD	Kurtosis	Skewness	H-lbqtest	H-chi²
1930–1939	−0.0196	2.0232	8.0295	0.298	1	1
1940–1949	0.0105	0.82	12.1952	−1.1305	1	1
1950–1959	0.0487	0.671	10.1739	−0.9169	1	1
1960–1969	0.0067	0.6533	8.5218	−0.0481	1	1
1970–1979	0.0006	0.9263	4.7826	0.2742	1	1
1980–1989	0.0487	1.1561	103.5361	−4.3501	1	0
1990–1999	0.0553	0.892	8.1889	−0.4086	1	1
2000–2008	−0.0087	1.2907	11.5553	0.0038	1	1
1928–2008	0.0175	1.1602	28.44	−0.6147	1	0

3.5 CONCLUSIONS

The statistical investigation of the DJIA daily logarithmic returns (r) show that r is not following a normal distribution in strict sense. While for the whole interval 1928–2000 the market is efficient overall, during particular decades there are deviations, as shown in the significant autocorrelations of returns. These deviations might be too small for practical purposes (trading), but they signal notable anomalies from the EMH hypothesis. EMH can be viewed as EMA (efficient market approximation), and the degree of approximation can be measured statistically for particular time periods.

Similar conclusions could be formulated describing new markets. Some studies reflect similar results for the Romanian financial market, describing financial data series as more asymmetric, and with kurtosis larger than 3. As other studies have indicated, EMH can be violated to some degrees in emerging markets, small markets, small companies stocks, and during periods of bubbles and crashes (Todea and Zoicaş-Ienciu, 2005; Todea, 2006; Săvoiu and Andronache, 2009). For very active markets as reflected by the DJIA index, the departure from EMH is small, yet interesting for further analysis of the effects of known anomalous periods such as crashes and bubbles.

ACKNOWLEDGMENTS

The authors acknowledge the use of DJIA index data.

REFERENCES

Bachelier, L., 1900. Théorie de la Spéculation. Annales Scientifiques de l'Ecole Normale Superieure, III 17, 21–86.

Beechey, M., Gruen, D. and Vickery, J., 2000. The efficient market hypothesis: A survey. Research Discussion Paper, Reserve Bank of Australia, Economic Research Department, Sydney.

Blakey, P., 2006. The efficient market approximation. IEEE Microwave Magazine 7 (1), 28–31.

Campbell, J.Y., Lo, A.W., MacKinlay, A.C., 1997. The Econometrics of Financial Markets. Princeton University Press, Princeton, NJ.

Fama, E.F., Fisher, L., Jensen, M, Roll, R., 1969. The adjustment of stock prices to new information. Int. Econ. Rev. 10 (1), 1–21.

Fama, E.F., 1970. Efficient capital markets: a review of theory and empirical work. J. Finan. 25 (2), 383–417.

Fama, E.F., 1991. Efficient capital markets. J. of Finan., 46 (5), 1575–1617.

Grossman, S.J., 1976. On the efficiency of competitive stock markets where traders have diverse information. J. Finan. 31 (2), 573–585.

Grossman, S.J., Stiglitz, J.E., 1980. On the impossibility of informationally efficient markets. Am. Econ. Rev. 70 (3), 393–408.

Jensen, M.C., 1978. Some anomalous evidence regarding market efficiency. J. Finan. Econ. 6 (2–3), 95–101.

Latham, M., 1985. Defining Capital Market Efficiency. Finance working paper 150 Institute for Business and Economic Research, University of California, Berkeley.

Lo, A.W., MacKinlay, A.C., 1999. A Non-Random Walk Down Wall Street. Princeton University Press, Princeton, NJ.

Malkiel, B., 1992. Efficient market hypothesis. In: Newman, P., Milgate, M., Eatwell, J. (Eds.), New Palgrave Dictionary of Money and Finance. Macmillan, London.

Rosenberg, J.M., 1982. Inside The Wall Street Journal, The Power and the History of Dow Jones & Company and America's Most Influential Newspaper. Macmillan, New York, NY.

Rubinstein, M., 1975. Securities market efficiency in an arrow-debreu economy. Am. Econ. Rev. 65 (5), 812–824.

Samuelson, P.A., 1973. Proof that properly discounted present values of assets vibrate randomly. Bell J. Econ. Manage. Sci. 4 (2), 369–374.

Săvoiu, G., Andronache, C., 2009. On the efficiency of financial markets. The 33rd Annual ARA Congress—American Romanian Academy of Arts and Sciences, Proceedings, vol. I, Polytechnic International Press, Montreal, Quebec, pp. 214–218.

Todea, A., 2006. La performance des methodes d'analyse technique sur le marche Roumain des actions: le cas des moyennes mobiles. Studia Oeconomica 51 (1), 75–86.

Todea, A., Zoicaş-Ienciu, A., 2005. Random and non-random walks in the romanian stock market. In: Poloucek, S., Stavarek, D. (Eds.), Future of Banking after the Year 2000 in the World and in the Czech Republic (Volume X – Finance and Banking). Silesian University, Karvina, pp. 634–646. < http://www.opf.slu.cz/kfi/icfb/2005/proceedings/2005_p04.pdf > Last (accessed 02 10 2012.).

Zhang, Y-C., 1999. Toward a theory of marginally efficient markets. Phys. A: Stat. Mech. Appl. 269 (1), 30–44.

CHAPTER 4

Nonlinear Mechanisms of Generating Power Laws in Socioeconomic Systems

Mircea Gligor

National College "Roman Vodă", Roman, Neamţ, Romania

4.1 WHAT ARE THE POWER LAWS?

4.2 ARE THE STOCK MARKET RETURNS GAUSSIAN DISTRIBUTED?

4.3 SOMETHING ABOUT THE RICHEST PEOPLE IN THE WORLD

4.4 FROM THE SMALLEST TOWNS TO THE LARGEST CITIES

4.5 WHAT MARK HAVE YOU GOT TODAY?

4.6 CONCLUSION

REFERENCES

4.1 WHAT ARE THE POWER LAWS?

Many of the things measured by the scientists have a typical size or "scale"—a typical value around which individual measurements are centered. The simplest examples are the heights of human beings, the speed of the cars (Newman, 2005), and the age of the pupils graduating the high school. In Figure 4.1 is plotted the distribution for this last example. We can easily see that almost all school leavers are 19 years old. There are no school leavers of 5 or 30 years old. The distribution is strongly centered on the value of 19 years old.

But not all things that we measure are peaked around a typical value. Some vary over an enormous dynamic range, sometimes many orders of magnitude. A classic example of this type of behavior is the population of towns and cities (Gligor and Gligor, 2008). The largest population of any city in Romania is 2,054,000 for Bucharest, as of the most recent (2003) census. The town with the smallest population is harder to find, since it depends on what we call a town. Using the

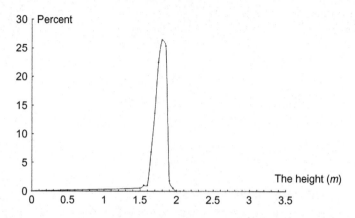

Figure 4.1 Histogram of ages of the pupils ending high school in a sample of 205 school leavers.

criteria from the Romanian Statistical Yearbook, we can pin up one of the smallest Romanian towns, i.e., Abrud, with 6,852. Whichever way we define a "small town," the ratio of largest to smallest population is about 300. This ratio is much larger for the countries that include extremely large cities. For instance, in the United States, the ratio is at least 150,000.

Clearly this is quite different from what we saw for age of the school leavers. The difference is clearly revealed when we look at the so-called Zipf plot: the populations n are set in decreasing order; next we define the rank of the city in a hierarchy, i.e., the largest city has $R = 1$; the second largest $R = 2$, etc.; then we plot n as a function of R. The diagram is highly left skewed, meaning that while the bulk of the distribution occurs for small sizes, there are a small number of cities with population much higher than the typical value. This left-skewed form is not itself very surprising. Given that we know there is a large dynamic range from the smallest to the largest city sizes, we can immediately deduce that there can only be a small number of very large cities.

What is surprising on the other hand is the right panel of Figure 4.2, which shows the histogram of city sizes again, but this time replotted with logarithmic horizontal and vertical axes. Now a remarkable pattern emerges: the diagram, when plotted in this fashion, follows quite closely a straight line. This observation seems first to have been made by Zipf (1949). What does it mean? Let $p(x)\mathrm{d}x$ be the

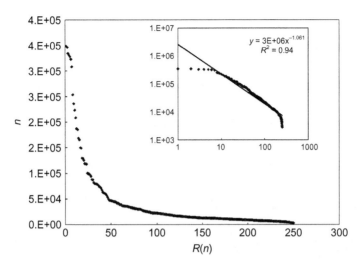

Figure 4.2 The urban population distribution for 256 Romanian cities and towns (the "Zipf plot"). Inset: the same distribution in log–log plot.

fraction of cities with population between x and $x + dx$. If the histogram is a straight line on log–log scales, then

$$\ln p(x) = -\alpha \ln x + C \tag{4.1}$$

where α and C are constants. (The minus sign is optional, but convenient since the slope of the line in the inset of Figure 4.2 is clearly negative.) Taking the exponential of both sides, this is equivalent to

$$p(x) = Cx^{-\alpha} \tag{4.2}$$

Distributions of the form (4.1) are said to follow a *power law*. The constant α is called the exponent of the power law. The constant C is mostly uninteresting; once α is fixed, it is determined by the requirement of normalization to 1.

Power law distributions occur in an extraordinarily diverse range of phenomena. Table 4.1 pointed out various phenomena from nature and society where such distributions occur.

A power law distribution is also sometimes called a *scale-free* distribution. That is because a power law is the only distribution that *is the same whatever scale we look at it on.* Suppose we have some

Table 4.1 Power Law Distributions in Nature and Society	
Quantity	**The Exponent α**
Magnitude of earthquakes (Bak, 1997)	3.04
Diameter of moon craters (Neukum and Ivanov, 1994)	3.14
Intensity of solar flares (Lu and Hamilton, 1991)	1.83
Urban area of the cities (Gligor and Gligor, 2008)	1.07
Population of the cities (Gligor and Gligor, 2008)	1.06
Relaxation of birth rate indices (Gligor and Ignat, 2001)	1.13–1.21
Income and wealth distribution (Gligor, 2005)	2.7–2.9
Intensity of wars (Roberts and Turcotte, 1998)	1.80
Magnitude of economic recessions (Ormerod and Mounfield, 2001)	1.28–1.36
Stock market indices (Gligor, 2004)	1.5–1.7
Number of hits on web sites (Adamic and Huberman, 2000)	2.40
E-mails received (Ebel et al., 2002)	2.22
Frequency of use of words (Zipf, 1949)	2.20
Frequency of family names (Miyazima et al., 2000)	1.94
Distribution of pupils marks (Gligor and Ignat, 2003)	1.3

probability distribution $p(x)$ for a quantity x and suppose we discover that it satisfies the property:

$$p(bx) = g(b)p(x) \qquad (4.3)$$

for any b. That is, if we increase the scale or units by which we measure x by a factor of b, the shape of the distribution $p(x)$ is unchanged, except for an overall multiplicative constant. This scale-free property is certainly not true of most distributions (it is not true for instance of the exponential distribution). In fact, as can be easily shown, it is only true of one type of distribution—the power law.

In the following sections, we analyze some nonlinear mechanisms that can explain the power law distributions occurrence in finance (Section 4.2), macroeconomics (Section 4.3), urbanism (Section 4.4), and psychology/learning theory (Section 4.5).

4.2 ARE THE STOCK MARKET RETURNS GAUSSIAN DISTRIBUTED?

George Soros pointed out in "Alchemy of finance" (Soros, 2003) the inadequacy and inefficiency of existing theories describing the behavior of stock prices. Until the last decade, theoretical economics was dominated by pure mathematics and was characterized by a ridiculous lemma/theorem style, with little effort being made to compare theoretical predictions with "experiment" (e.g., prices from real stock markets) and by the fact that the bulk of papers were inaccessible and of no interest to "experimentalists," i.e., practitioners in the field. Pure mathematics has nevertheless made important contributions to economics through the game theory approach (the concept of Nash equilibrium, in which no player can improve on his/her strategy and the perfect rationality of all players is assumed) and through the phenomenology of stock price fluctuations that are postulated to be Gaussian and subject to the "efficient market hypothesis"—that all correlations are arbitraged away. There are however many observations which disagree with these suppositions: first of all, short-term fluctuations are not Gaussian; second, price increments are correlated (the magnitude of price fluctuations has extended temporal correlations); and third, strategies used by traders are also correlated, as a consequence of the herd effect.

Mandelbrot (1963) was the first to emphasize the idea of comparing the distributional properties of price changes on different timescales. The idea behind Mandelbrot's approach is that of scale invariance: the distribution p_T of price changes on a timescale T may be obtained from that of a shorter timescale $\tau < T$ by an appropriate rescaling of the variable:

$$P_T(x) = \frac{1}{\lambda} P_\tau \left(\frac{x}{\lambda} \right) \tag{4.4}$$

where $\lambda = (T/\tau)^H$ and H is the self-similarity (Hurst) exponent. In this way, the concepts of scale invariance and scaling behavior were firstly applied outside their traditional domains of application in the physical sciences.

In Figure 4.3, one can see the presence of the large fluctuations outside the Gaussian models. Moreover, in Figure 4.4, the semilogarithmic plot of the probability density function allow us to see that while the central part of the distribution may be approximate to Gaussian, the asymptotic branches ("the fat tails") are well fitted to power laws (Gligor, 2004).

We now introduce a model to indicate how the power law tails may be derived from the Gaussian distribution by accounting for the random amplification process. Let $p(y)$ denote the basic distribution written in terms of dimensionless quantity $y = x/<x>$, $<x>$ being the mean value of the observed x if the tail of the distribution is neglected. With a small probability λ, suppose that in the new amplifier class one has the same distribution function p that is natural for the process but that $<x>$ is amplified to $N<x>$. Thus, $p(y)dy$ becomes $(1/N)p(y/N)dy$. In the second stage of amplification, which we postulate to occur with a probability λ^2, the

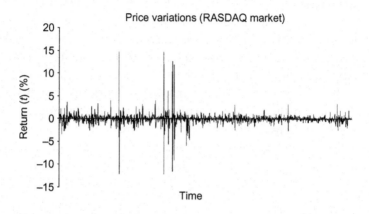

Figure 4.3 The daily changes (%) of the financial index RASDAQ-C from July 31, 1998 to April 17, 2002. One can see large fluctuations (over 10%) well outside the Gaussian regime.

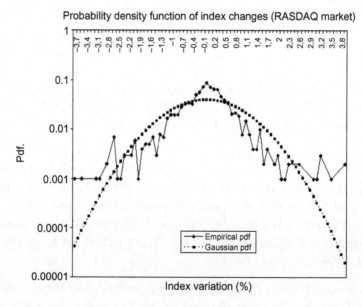

Figure 4.4 The probability density function of stock returns in semilogarithmic plot. The fat tails of the empirical distribution are in striking contrast with the predictions of Gaussian models.

mean value of x becomes $N^2 <x>$. The new distribution $P(y)$ that allows for the possibility of continuing levels of amplification is

$$P(y) = (1 - \lambda)\left[p(y) + \frac{\lambda}{N}p(y/N) + \frac{\lambda^2}{N^2}p(y/N^2) + \cdots\right] \qquad (4.5)$$

where the factor $(1 - \lambda)$ is introduced to ensure the normalization of $P(y)$, i.e., $\int P(y)dy = 1$.

It is easy to see that by replacing $y \rightarrow y/N$ in Eq. (4.2), $P(y)$ is given by the scaling equation:

$$P(y) = \frac{\lambda}{N}P\left(\frac{y}{N}\right) + (1 - \lambda)p(y) \qquad (4.6)$$

Although the determination of the complete solution of this inhomogeneous equation is rather complex, it is easy to obtain the asymptotic properties of $P(y)$. When $y \rightarrow \infty$, then $p(y) \rightarrow 0$. Let us suppose this decay is faster than that of $P(y)$. Then the asymptotic form of $P(y)$ is determined by the simpler equation:

$$P(y) = \frac{\lambda}{N}P\left(\frac{y}{N}\right) \qquad (4.7)$$

Checking a solution of the form $P(y) = (\text{const.})y^{-1-\alpha}$, the direct substitution yields

$$\alpha = \frac{\log(1/\lambda)}{\log N} \qquad (4.8)$$

Thus, the power law exponent appears as a fractional dimension, as we have found explicitly starting from the empirical data.

4.3 SOMETHING ABOUT THE RICHEST PEOPLE IN THE WORLD

The study of income distribution has a long history. It is a well-known fact that the individual wealth is a very broadly distributed quantity among the population. More than a century ago, the Italian sociologist V. Pareto studied the distribution of personal incomes for the purpose of characterizing a whole country's economic status (Pareto, 1897). The cumulative distribution of wealth is often described by "Pareto" tails, which decay as a power law for large wealth:

$$P_>(W) \sim \left(\frac{W_0}{W}\right)^{\mu} \qquad (4.9)$$

where $P_>(W)$ is the probability to find an agent with wealth greater than W and μ is a certain exponent of order 1.7–1.9, both for individual wealth and incomes. In 1921, Gini checked the same statistics and reported that power laws actually hold, but the values of the exponents vary from country to country. Mandelbrot (1982) claimed that the wealth distribution turned out to be a classic example of fractal distribution and proposed a "weak Pareto law" applicable only asymptotically to the high incomes. Montroll (1987) analyzed the United State's personal income data for the year 1935/1936 and found that the top 1% of incomes follow a power law with an exponent 1.63, while the rest, who are expected to be salaried, follow a lognormal distribution. Most recent studies have confirmed the fact that in the most countries, over 90% of the total wealth is owned by maximum 5% of the population, while for the rest, the wealth distribution is well fitted either the exponential or the lognormal distribution.

For example, let us consider the wealth distribution in Japan. The tabulated data of the income of 6,224,249 individuals in Japan in fiscal year 1998 are available on the web (The Japanese Tax Administration, Burniaux et al., 2008).

The cumulative distribution (the "Pareto diagram") is plotted in Figure 4.5 in log–log scale. One can easily see that for the great majority of the population, the diagram is well fitted by the parabola (this meaning a lognormal distribution), while the asymptotic tail is well fitted by a straight line (the power law).

A simple nonlinear mechanism that can explain this result was elaborated in 2000 (Solomon, 2000).

In order to obtain power law probability distributions $P(w) \sim w^{-1-\alpha}$ with $\alpha > 0$, we consider the multiplicative process:

$$w(t + 1) = \lambda(t)w(t) \qquad (4.10)$$

where the random variables $\lambda(t)$ are extracted from a fixed probability distribution $\Pi(\lambda)$ with positive support. We take the logarithm on both sides of Eq. (4.10) and use the notations $\mu = \ln \lambda$ and $x = \ln w$:

$$x(t + 1) = \mu(t) + x(t) \qquad (4.11)$$

The variation of $w(t)$ is constrained by a lower bound (or barrier):

$$w(t) > w_{min} \qquad (4.12)$$

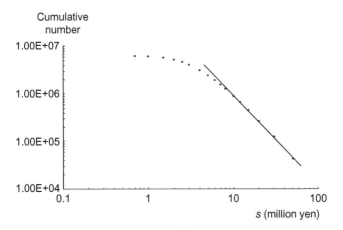

Figure 4.5 The Pareto diagram of wealth distribution in Japan in 1998.

In terms of $x(t) = \ln w(t)$, Eq. (4.11) becomes supplemented by the condition

$$x(t) > x_{min} = \ln w_{min} \tag{4.13}$$

Actually, the dynamics of $w(t)$, that consists at each time t in the updating Eq. (4.10)—or, equivalently, Eq. (4.11)—will be modified as follows: if $w(t+1) < w_{min}$ (or, equivalently, $x(t+1) < x_{min} \equiv \ln w_{min}$), then the updated value of the variable is $w(t+1) = w_{min}$, respectively $x(t+1) = x_{min}$. Equation (4.11) with the constraint equation (4.13) is one of the equations most frequently met in physics (the "the barometric problem") and has the solution

$$P(x) \sim \exp\left(-\frac{x}{kT}\right)$$

where x can be considered as the coordinate of a molecule in the earth gravitational field, the "earth surface" being x_{min}, k is the Boltzmann constant and T is the temperature.

Now we express the solution in w's terms using the new distributions:

$$\rho(\ln \lambda)d(\ln \lambda) = \prod(\lambda)d\lambda \tag{4.14a}$$

$$\Psi(\ln w)d(\ln w) = P(w)dw \tag{4.14b}$$

we get the solution:

$$P(w)dw \sim \exp\left(-\frac{\ln w}{kT}\right) d(\ln w)$$

i.e.,

$$P(w) \sim w^{-1-1/kT} \equiv w^{-1-\alpha} \tag{4.15}$$

One can easily see that the exponent α in the power law (4.6) is related to the average w:

$$\overline{w} = \frac{1}{N}\sum_i w_i \tag{4.16}$$

As well, α is related to w_{min} by means of a new relevant parameter, q, defined as

$$q = \frac{w_{min}}{\overline{w}} \tag{4.17}$$

Imposing the normalization:

$$\int_{w_{min}}^{\infty} P(w)dw = 1$$

we get the complete form of the probability density function:

$$P(w) = \alpha w_{min}^{\alpha} w^{-1-\alpha} \tag{4.18}$$

Imposing now that the averaged w is $\overline{w}$:

$$\int_{w_{min}}^{\infty} wP(w)dw = \overline{w}$$

we obtain the exponent of the power law:

$$\alpha = \frac{1}{1-q} \tag{4.19}$$

In Eq. (4.19), we have obviously $q > 0$ that it predicts $\alpha > 1$ according with the empirically observed exponents of wealth distribution.

4.4 FROM THE SMALLEST TOWNS TO THE LARGEST CITIES

Let us recall from the first section that the distribution of city and town populations is well fitted to a power law with the exponent close to unit (Figure 4.2). This result was firstly pointed out in Zipf's pioneering book (Zipf, 1949). If n_s is the number of cities having the population s, then

$$R(s) = \int_s^\infty n_x \, dx \qquad (4.20)$$

defines the rank of the city in a hierarchy, i.e., the largest city has $R = 1$; the second largest $R = 2$, etc. Zipf found that R is a function of s, which can be inverted as

$$s(R) \sim R^{-\gamma}, \quad \text{with } \gamma \approx 1 \qquad (4.21)$$

The first striking property of the above result (known today as "the Zipf's law for cities" is the scale invariance, reflecting an underlying fractal structure; the second consists in universality: the statistical datasets shows that the law is valid for many different societies and during various time periods.

The universality of the power law behavior suggests the possibility of study the urban system by tools that do not depend in an explicit way on the concrete nature of the interactions between its elementary constituents (Blank and Solomon, 2000; Gabaix, 1999; Marsili and Zhang, 1998).

We show below that Zipf's law can be easily derived by supposing that the development process is Markovian (Gligor and Gligor, 2008). It is well known that Markovian stochastic processes can be described by a master equation.

Let us consider N towns, and let s_i be the size of the ith city (expressed as number of citizens as well as units of urban area). The model is built using the general framework of the master equation. We assign the transition rates for the growth $\Psi_+(s_i)$ or decrease $\Psi_-(s_i)$ of the size s_i. In other words, $\Psi_+(s_i)$ is the probability that a new citizen arrives (or a new economic/residential unit-area location is created) in the city i in the time interval $(t, t + dt)$, so that $s_i \to s_i + 1$. Analogously, $\Psi_-(s_i)$ is the probability that one of the s citizens departs (or a unit-area location is left) in the same time interval, so that $s_i \to s_i - 1$.

We introduce now the average number $n(s, t)$ of cities of size s at time t, for a *given* N. The quantity $n(s, t)$ satisfies the master equation:

$$\frac{\partial n(s, t)}{\partial t} = \Psi_-(s + 1)n(s + 1,\ t) - \Psi_-(s)n(s,\ t)$$
$$+ \Psi_+(s - 1)n(s - 1, t) - \Psi_+(s)n(s,\ t) \tag{4.22}$$

The parameters of the model are the transition rates $\Psi_\pm(s_i)$.

If the total number of cities N is considered not to be constant, at least an additional parameter must be introduced, describing the probability that a citizen leaves the system (or a unit-area location is created outside of the system). However, the timescale of this process certainly exceeds the timescale of the statistical data recording for this model.

Thus, as in the study of the most interacting-agent systems, we are firstly interested in finding the stationary solution of the master equation, for which s and N are constant on average, and

$$\frac{\partial n(s,\ t)}{\partial t} = 0 \tag{4.23}$$

In this case $n(s,\ t) \equiv n(s)$, i.e., the quantities n and Ψ do not depend explicitly on the time. Equation (4.26) becomes

$$\Psi_-(s + 1)n(s + 1) - \Psi_-(s)n(s) + \Psi_+(s - 1)n(s - 1) - \Psi_+(s)n(s) = 0 \tag{4.24}$$

The simplest way to take into account the interactions among agents is assuming these interactions pairwise type, so that $\Psi \sim s^2$. This assumption simply means that all the s city units are in interaction with each other, displaying a fully connected social network. In the simplest way, choosing $\Psi_-(s) = k_1 s^2$ and respectively $\Psi_+(s) = k_2 s^2$, a straightforward calculus leads to

$$n(s) = C/s^2 \tag{4.25}$$

where $C = N/\sum_{i=1}^{N} s_i^{-2}$. Using the rank relation (4.24), one finds $R(s) \sim s^{-1}$ or inverting, $s(R) \sim 1/R$, i.e., the usual form of Zipf's law.

4.5 WHAT MARK HAVE YOU GOT TODAY?

A statistical study on various samples of marks in the high school has shown that their distribution on mark levels displays some universal features (Gligor and Ignat, 2003). Using items with multiple choices, the deviation from the Gaussian form becomes relevant and the distribution gets the shape of a stable Levy distribution with a fat tail that can be fitted to a power law. The symmetric Levy distribution is known analytically in the tails where

$$L(x) \to \frac{c^{\alpha}\Gamma(1 + \alpha)\sin(\pi\alpha/2)}{\pi|x|^{\alpha+1}}, \quad x \to \infty \tag{4.26}$$

where c is the scale factor and $\alpha < 2$ is the characteristic exponent. The tail of empirical distribution is well fitted to a power law with $1 + \alpha \cong 1.3$ (Figure 4.6).

In their famous paper published in 1987, Bak et al. introduced the term "self-organized criticality" in order to describe many natural phenomena that display fractal behavior, i.e., long-range correlations with power law decay over a wide range of length scales (Bak, Tang and Wiesenfeld, 1987).

The main idea of this approach is that the dynamics that gave rise to the robust power law correlations had not to involve any fine-tuning of parameters. The systems, under their natural evolution, are driven to a state at the boundary between the stable and unstable states. Such a state then shows long-range temporal–temporal fluctuations similar to those in equilibrium critical phenomena.

We propose a simple example of a system whose natural dynamics drives it toward, and then maintains it, at the edge of stability: a

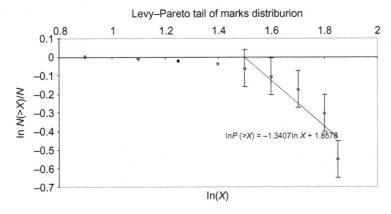

Figure 4.6 The logarithm of the number of marks greater than X vs. logarithm of X.

sandpile. For dry sand, one can characterize its macroscopic behavior in terms of an angle θ_c, called the angle of repose, which depends on the detailed structure (size, shapes, roughness, etc.) of the constituting grains. If we make a sandpile in which the local slope is smaller than θ_c everywhere, such a pile is stable. On such a pile (Figure 4.7), the addition of a small amount of sand will cause only a weak response. The empirical data (little diamonds) are here fitted to the straight line. The power law exponent is $1 + \alpha \cong 0.34$. Error bars are bootstrap 95% confidence intervals (Gligor and Ignat, 2003).

The addition of a small amount of sand to a configuration where the average slope is larger than θ_c will often result in an avalanche whose size is of the order of the system size. In a pile where the average slope is θ_c, the response to addition of sand is less predictable. It might cause almost no relaxation; it may cause avalanches of intermediate sizes, or a catastrophic avalanche that affects the entire system. Such a state is critical. If one builds the sandpile on a finite table by pouring it very slowly, the system is invariably driven toward its critical state and thus organizes itself into a critical state: it shows "self-organized criticality."

The steady state of this process is characterized by the following property: sand is being added to the system at a constant small rate, but it leaves the system in a very irregular manner, with long periods of apparent inactivity interspersed by events which may vary in size and which occur at unpredictable intervals.

A similar behavior is seen in earthquakes, where the buildup of stress due to tectonic motion of the continental plates is a slow steady process, but the release of stress occurs sporadically in bursts of various sizes.

We propose further a simple cellular automata model of sandpile growth. The model is defined on a lattice, which we take for simplicity

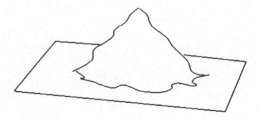

Figure 4.7 A small sandpile on a flat table.

to be the two-dimensional square lattice. There is a positive integer variable at each site of the lattice, called the height of the sandpile at that site. The system evolves in discrete time.

The rules of evolution are quite simple: at each time step a site is picked randomly, and its height z_i is increased by unity. If its height is then larger than a critical height $z_c = 4$, this site is said to be unstable. It relaxes by toppling whereby four sand grains leave the site, and each of the four neighboring sites gets one grain. If there is any unstable site remaining, it too is toppled. In case of toppling at a site at the boundary of the lattice, grains falling "outside" the lattice are removed from the system. This process continues until all sites are stable.

Then another site is picked randomly, its height increased, and so on. It is easy to show that this process must converge to a stable configuration in a finite number of time steps on any finite lattice using the diffusive nature of each relaxation step.

The following example illustrates the toppling rules. Let the lattice size be 4×4, and suppose at some time step the following configuration is reached:

2	3	1	4
4	4	2	1
4	2	1	1
3	1	3	2

We now add a grain of sand at a randomly selected site: let us say the site on the second row from the top and the second column from the left. Then it will reach height 5, become unstable, and topple to reach:

2	3	1	4
4	5	2	1
4	2	1	1
3	1	3	2

$\longrightarrow$

2	4	1	4
5	1	3	1
4	3	1	1
3	1	3	2

The further toppling results in

3	4	1	4
1	2	3	1
5	3	1	1
3	1	3	2

$\longrightarrow$

3	4	1	4
2	2	3	1
1	4	1	1
4	1	3	2

In the last configuration, all sites are stable. One speaks in this case of an event of size $s = 3$, since there were three topplings. Other measure of event size could be the number of time steps needed (in this case also three). Suppose that one measures the relative frequency of event sizes. A long power law tail would characterize a typical distribution, with an eventual cutoff determined by the system size (Figure 4.8).

The plot in Figure 4.8 is constructed starting from a numerical simulation (Gligor and Ignat, 2004) that has led to a relative small number (361) of events. Nonetheless, the critical exponent is found to be in good agreement with the one calculated from the empirical data referring to the marks distribution. Note that some stronger simulations using over 10^6 events and running on lattices up to 100×100 (Pakzuski et al., 1996) have approximately led to the same critical exponent, in agreement with the one that we extracted from the data referring to the distribution of marks in school (Gligor and Ignat, 2003).

The data (little diamonds) were generated by simulation with 361 events on a 4×4 lattice. The continuous line is the power law fitting of exponent $1 + \alpha = 1.394$.

Thereby, one may claim that the scaling properties reflected by the power law distribution are in relation with the learning mechanisms and thus they can be incorporated in the general framework of the learning theory. The result might be a starting point in investigating the brain activity (in particular the learning processes) as a succession of avalanches of various orders of size, in a system running in a self-organized critical state.

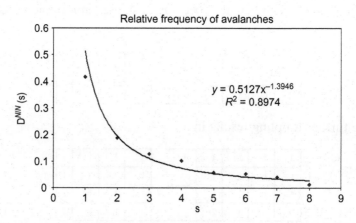

Figure 4.8 The relative frequency of events as a function of their sizes (s).

4.6 CONCLUSION

The key concept discussed in the present paper was "system composed of many interacting agents." Such a system, topically called "complex system," displays some emergent properties that cannot be reduced to the sum of their individual features; the nonlinearity, on one hand, and the interaction, on the other hand, make the difference.

The importance of power laws is, by far, more than academic: the power laws constitute veritable bridges between the microscopic (individual) laws and the macroscopic (collective) behavior. The phenomena described by power laws cover all the dynamic scales, and it is to a large extent independent of the microscopic details of the system. In other words, the macroscopic complexity is not a direct consequence of the microscopic dynamics but rather an effect of self-organization, interconnections, and feedback of the interacting agents.

In conclusion, the unifying and predictive capacity of the power laws makes them the subject of study in various fields, from quantum and statistical physics to economics, sociology, ecology, and psychology. Taking into account the results obtained so far, we can claim that the power laws emergence is a *sine qua non* condition for generating the macroscopic world starting from the microscopic laws. This fact might be considered, by itself, a new fundamental law of nature.

REFERENCES

Adamic, L.A., Huberman, B.A., 2000. The nature of markets in the World Wide Web. Q. J. Electron. Comm. 1, 512–517.

Bak, P., 1997. How Nature Works: The Science of Self-Organized Criticality. Oxford University Press, Oxford.

Bak, P., Tang, C., Wiesenfeld, K., 1987. Self-organized criticality: an explanation of $1/f$ noise. Phys. Rev. Lett. 59, 381–384.

Blank, A., Solomon, S., 2000. Power laws in cities population, financial markets and internet sites (scaling in systems with a variable number of components). Phys. A 287, 279–288.

Burniaux et al., 2008. Income distribution and poverty in selected OECD countries: Economics department working papers no. 189. Available at: < http://www.oecd.org/eco/productivityan-dlongtermgrowth/1864447.pdf > Last (accessed 02.10.2012).

Ebel, H., Mielsch, L.-I., Bornholdt, S., 2002. Scale-free topology of e-mail networks. Phys. Rev. E 66, 035103.

Gabaix, X., 1999. Zipf's law for cities: an explanation. Q. J. Econ. 114, 739–767.

Gini, C., 1921. Measurement of inequality of incomes. Economic Journal, 31 (121), 124–126.

Gligor, L., Gligor, M., 2008. The fractal city theory revisited: new empirical evidence from the distribution of romanian cities and towns. Nonlinear Dyn. Psychol. Life Sci. 12 (1), 15–28.

Gligor, M., 2004. An empirical study on the statistical properties of Romanian emerging stock market RASDAQ. Int. J. Theor. Appl. Finan. 7, 723–739.

Gligor, M., 2005. Extremum criteria for non-equilibrium states of dissipative macroeconomic systems. In: Sieniutycz, S., Farkas, H. (Eds.), Variational and Extremum Principles in Macroscopic Systems. Elsevier Ltd., The Boulevard, Langford Lane Kidlington, Oxford, OX5 IGB UK, pp. 717–734.

Gligor, M., Ignat, M., 2001. Some demographic crashes seen as phase transitions. Phys. A 301, 535–544.

Gligor, M., Ignat, M., 2003. Scaling in the distribution of marks in high school. Fractals 11, 363–369.

Gligor M., Ignat, M., 2004. Self-organized criticality and the learning process. The XXXIIIth National Conference Physics and the Modern Educational Technologies, May 13–15, Iaşi.

Lu, E.T., Hamilton, R.J., 1991. Avalanches of the distribution of solar flares. Astrophys. J. 380, 89–92.

Mandelbrot, B.B., 1963. The variation of certain speculative prices. J. Business 36, 394–419.

Mandelbrot, B.B., 1982. The Fractal Geometry of Nature. Freeman, San Francisco, CA.

Marsili, M., Zhang, Y.C., 1998. Interacting individuals leading to Zipf's law. Phys. Rev. Lett. 80, 2741–2744.

Miyazima, S., et al., 2000. Power-law distribution of family names in Japanese societies. Phys. A 278, 282–288.

Montroll, E.W., 1987. On the dynamics and evolution of some socio-technical systems. In: West, B.J. (Ed.), Bulletin (New Series) of the American Mathematical Society, 16. pp. 1–54, Publisher: American Mathematical Society, Providence, Rhode Island, USA. E-print: http://www.ams.org/journals/bull/1987-16-01/S0273-0979-1987-15462-0/S0273-0979-1987-15462-0.pdf Last accessed: The 2nd of October, 2012.

Neukum, G., Ivanov, B.A., 1994. Crater size distributions and impact probabilities on earth from lunar, terrestial planet, and asteroid cratering data. In: Gehrels, T. (Ed.), Hazards Due to Comets and Asteroids. University of Arizona Press, Tucson, AZ, pp. 359–416.

Newman, M.E.J., 2005. Power laws, Pareto distributions and Zipf's law. Contemp. Phys. 46, 323–351.

Ormerod, P., Mounfield, C., 2001. Power law distribution of the duration and magnitude of recessions in capitalist economies: breakdown of scaling. Phys. A 293, 573–582.

Pakzuski, M., Maslov, S., Bak, P., 1996. Avalanche dynamics in evolution, growth, and depinning models. Phys. Rev. E 53, 414–443.

Pareto, V., 1897. Le Cours d' Economie Politique. Macmillan, London.

Roberts, D.C., Turcotte, D.L., 1998. Fractality and selforganized criticality of wars. Fractals 6, 351–357.

Solomon, S., 2000. Generalized Lotka Volterra (GLV) models of stock markets. In: Ballot, G., Weisbuch, G (Eds.), Applications of Simulation to Social Sciences. Hermes Science Publications, pp. 301–322. , < http://xxx.lanl.gov/abs/cond-mat/9901250 > Last (accessed 02.10.2012.).

Soros, G., 2003. The Alchemy of Finance (Wiley Investment Classics). John Wiley & Sons, Inc., Hoboken, NJ 07030-5774, USA, pp. 67–71.

Zipf, G.K., 1949. Human Behaviour and the Principle of Least Effort. Addison-Wesley, Reading, MA.

CHAPTER 5

Was It Possible to Forecast the Credit Crunch? (Monte Carlo Simulation of Integrated Market and Credit Risk)

Aretina-Magdalena David-Pearson
Bucharest Polytechnic University, Faculty of Applied Sciences, Romania

5.1 INTRODUCTION

Financial institutions have taken steps to measure risk in an effort to satisfy regulatory requirements and avoid losses. Their efforts have primarily focused on the calculation of market risk for a particular portfolio. This approach, however, does not capture the interaction of the different types of risk that a firm faces, the marginal impact of a trade across the institution, or the effects of offsetting market moves.

The need for an integrated approach for market and credit risk emerged with the beginning of the twenty-first century and became evident with the burst of the credit crunch in 2007. Combining together market and credit risk is assembling the overall financial risk of the portfolio in quantities easier to manage. The complexity of the actual financial markets points to the increased need of being able to integrate both credit and market risk in a unitary way.

5.2 INTRODUCTION TO FINANCIAL THEORY

Market risk is defined as the risk that a financial position changes its value due to the change of an underlying market risk factor, like a stock price, an exchange rate, or an interest rate.

Credit risk is defined as the risk that an obligor will not be able to meet its financial obligations toward its creditors. Under this definition, default is the only credit event. The weaker definition of credit risk is based on market perception. This definition implies that obligors will face credit risk even if they do not fail their financial obligations yet but the market perceives they might fail in the future. This is known as the mark-to-market definition of credit risk and gives rise to migration as well as default as possible credit events. Perception of financial distress gives rise to credit downgrade.

Value at risk (VaR) is an estimate of the maximum loss that can occur for a portfolio under market conditions and within a given confidence level over a certain period of time. *VaR* has become an increasingly important measure for strategic risk management. Recent market events, including the Asian crisis and market collapse in Russia, have underscored the importance of complementing *VaR* analysis with a comprehensive stress-testing program.

New models and techniques for the trading book portfolios are needed to calculate the market risk. Credit is now playing a role in pricing—credit spreads are key factors for pricing many instruments. Specific risks arise from credit quality changes in the market place. Thus the changes in credit risk affect market prices through spreads or credit ratings, infringing on the purity of the market risk estimation.

As credit markets expand and deepen, information such as spreads and downgrades contributes directly, in an increased manner, to the positions' valuation in the corporate bond portfolios. Similarly, because market rates drive the value of fixed rate bonds, counterparty credit risk can only be assessed in such portfolios when exposures are evaluated under a variety of market conditions. Hence market risk factors are fundamental for a correct measure of credit risk. Credit risk modeling is one of the top priorities in risk management and finance.

Common practice still treats market and credit risk separately. When measuring market risk, credit risk is commonly not taken into account; when measuring portfolio credit risk (PCR), the market is assumed to be constant. The two risks are then "added" in *ad hoc* ways, resulting in an incomplete picture of risk. In this study, the focus is on integrated credit and market risk.

There are two categories of credit risk measurement models: *counterparty credit exposure* models and *PCR* models.

Counterparty credit exposure is the economic loss that will be incurred on all outstanding transactions if a counterparty defaults unadjusted by possible future recoveries. Counterparty exposure models measure and aggregate the exposures of all transactions with a given counterparty. They do not attempt to capture portfolio effects, such as the correlation between counterparty defaults.

PCR models measure credit capital and are specifically designed to capture portfolio effects, specifically obligor correlations. It accounts for the benefits of diversification. With diversification, the risk of the portfolio is different from the sum of risk across counterparties. Correlations allow a financial institution to diversify their portfolios and manage credit risk in an optimal way. However, PCR models either fix market risk factors to account for credit risk or fix credit risk drivers to account for market risk.

In this study, the *portfolio credit risk engine (PCRE)* is used, which is the first production solution for integrated market and credit risk, based on conditional probabilities of default.

Mark to future (MtF) is a concept where all financial instruments are valued across multiple scenarios (developed on the underlying market and credit) risk factors and across the time steps of interest. In the *MtF* concept, the calculations for pricing the instruments and *VaR* estimates are retrieved to deliver any combinations of results according to financial instrument, scenario, and time step.

If portfolio positions depend simultaneously on market and credit risk factors, the nature of the risk assessment problem changes. If market and credit risks are calculated separately, this is based on a wrong portfolio valuation and leads to a wrong assessment of true portfolio risk.

A comprehensive framework requires the full integration of market and credit risk. It is mandatory to model the correlations between changes of the credit quality of the debtors and changes of market risk factors. By combining an MtF framework of counterparty exposures (Aziz and Charupat, 1998) and a conditional default probability framework (Gordy, 1998), one minimizes the number of scenarios where expensive portfolio valuations are calculated and can apply advanced Monte Carlo or analytical techniques that take advantage of the problem structure. This integrated structure has the advantage of explicitly defining the joint evolution of market risk factors and credit drivers. Market factors drive the prices of securities and credit drivers are macroeconomic factors that drive the creditworthiness of obligors in the portfolio.

5.3 METHODOLOGY OF CREDIT DRIVERS

As described in Kane, Krupp and Macki (2001), the credit driver's correlation infrastructure and data allow the impact of correlated counterparty defaults and migrations to be incorporated in to the measurement of enterprise PCR and credit risk capital. This infrastructure helps in the computation of credit losses—based on global empirical data for correlations. Correlated counterparty defaults and migrations in the PCR model are simulated based on the evolution of the counterparty *creditworthiness index (CWI)*.

CWI comprises systemic credit risk arising from movements in the credit drivers that are common to the counterparties in the portfolio and idiosyncratic (unsystematic) risk that is specific to a particular counterparty in the portfolio. For a portfolio of J obligors, the credit quality of an obligor is modeled through a multifactor CWI, Y_j, described in following equation:

$$Y_j(t) = \sum_{i=1}^{N} \beta_{ij} Z_i(t) + \alpha_j \varepsilon_j \tag{5.1}$$

In the above equation, the systemic credit risk component of the CWI for each obligor j, $j = 1, 2, \ldots, J$, is assumed to be driven by credit drivers Z_i, $i = 1, 2, \ldots, N$. Each credit driver Z_i represents the country and industry sector affiliation of obligor j.

The sensitivities vector can then be written as $(\beta_{1j}, \beta_{2j}, \ldots, \beta_{ij}, \ldots, \beta_{Ij})$, where β_{ij} is the sensitivity of obligor j to credit driver i.

The second term in Eq. (5.1) represents the obligor-specific, idiosyncratic risk component. The higher the sensitivities of an obligor to a credit driver or a set of them, the higher its systemic risk and the lower its idiosyncratic risk will be.

$$\alpha_j = \sqrt{1 - \sum_{i=1}^{N} \beta_{ij}^2} = \sqrt{1 - R_j^2} \qquad (5.2)$$

where R_j^2 is the proportion of variance of Y_j explained by the credit drivers and ε_j in Eq. (5.1) are independent standard normal variables. One refers to α_j as the specific weight for counterparty j.

In the above framework, correlation between any two obligors' credit quality is governed by the correlations among the risk drivers in their CWIs. This is represented by the joint *variance covariance (VCV)* matrix.

The covariance between the CWIs Y_l and Y_j of any two obligors can be written as

$$\mathrm{cov}(Y_l, Y_j) = \sum_{i=1}^{N} \sum_{k=1,k \neq i}^{N} \beta_{il}\beta_{kj}\mathrm{Cov}(Z_i, Z_k) + \sum_{i=1}^{N} \beta_{il}\beta_{ij}\mathrm{Var}(Z_i) \qquad (5.3)$$

Under a multifactor CWI model, the sensitivity β_{ij} must be estimated for a set of credit drivers appropriate for a given counterparty.

We have also assumed that when all except one of the sensitivities to the credit drivers equal zero, the sensitivities vector for obligor j is with systemic impact of only one credit driver, namely, Z_j. In this case, although the systemic risk in each obligor CWI is determined by a single index, the model still captures the diversification effect across different region–industry sector pairs through the different indexes serving as proxies for such pairs. At the same time, correlated defaults are captured to the extent that the indexes themselves are correlated. Once we estimate the CWI for each counterparty in the credit portfolio, the occurrence of default or migration in each one of a set of Monte Carlo scenarios on the credit drivers can be simulated. Also, given other information such as counterparty exposures, recovery rates, length of planning, and horizon, the portfolio loss distribution is computed where the losses now incorporate the impact of correlated defaults and migrations. A prespecified loss percentile, e.g., 99.95th, of the loss distribution then signifies the credit risk economic capital for the portfolio.

In theory, the above process requires prior knowledge of the specific counterparty names in the credit portfolio in order to compute the credit loss distribution. But in the above infrastructure and framework, however, one eliminates this requirement of prior knowledge of specific names. Instead, one estimate CWIs for representative counterparties on to which specific counterparties in any given portfolio may be mapped, based on a set of well-defined criteria. We shall now describe the CWI estimation methodology.

As we have seen above, we assume a multifactor CWI model wherein a given counterparty has a single credit driver associated to it. Given our objective of arriving at an infrastructure whereby credit losses can be computed for an arbitrary credit portfolio that may contain counterparties from any part of the globe, we have first identified a comprehensive set of credit drivers with global coverage. In addition to providing comprehensive coverage, the selected credit drivers also capture the characteristics of the economic and credit environment of a given region as well as the characteristics of the particular industry within which a given counterparty is operating. We have divided the globe into six regions, as shown in Table 5.1. Further, within each region, we have utilized the *global industry classification system (GICS)* for 10 industry sectors in order to account for their distinct credit risk characteristics. We have a total of 60 region—sector combinations, to cover a global credit portfolio, as shown in Table 5.1.

For each region—sector combination, we designate the corresponding Dow—Jones region—sector index as the credit driver for counterparty affiliated to that region and sector.

In order to estimate the sensitivity of a representative counterparty for a given region—sector combination to the region—sector index as the credit driver, we have carried out the following steps. First, we have collected the time series data on all fixed rate corporate bonds in a given region and sector and the time series data on the corresponding region—sector indexes. Next, we use the average of the estimated *R-squares* for the bonds in the given region—sector combination obtained when the individual fixed income returns are regressed on the index returns, which are then available for use as the *R-square* for a representative counterparty in a given region—sector combination.

Table 5.1 Description of Credit Drivers: Six Regions and 10 Sectors for 60 Credit Drivers	
Regions	Sectors
1. United States 2. Americas (ex-United States) 3. United Kingdom 4. Europe (ex-United Kingdom) 5. Japan 6. Asia-Pacific (ex-Japan)	1. Energy 2. Basic Materials 3. Industrials 4. Consumer Cyclical 5. Consumer Non-Cyclical 6. Health Care 7. Financials 8. Information Technology 9. Telecommunication Services 10. Utilities

Once *R-square* estimates are available for a representative counterparty for every region—sector combination, Eq. (5.2) can be used to estimate the sensitivity and the specific risk for any counterparty in a given region and sector therein.

5.4 DATA DESCRIPTION AND ANALYSIS

A portfolio of fixed rate bonds is taken. It comprises of 140 portfolios denominated in different currencies with different credit ratings assigned to each one of them. USD has been considered as the base currency.

A risk factor can be any observable economic variable whose value, or change in value, may result in change in the value of the portfolio. The set of all risk factors and their values provide with an "economic snapshot" under which the portfolio is evaluated during simulation. A scenario represents a possible future economic situation. A scenario set is often interpreted as the complete set of possible future economic situations. It is a list of risk factors and their values at one or more points in the future. Scenarios are the language of risk. A sample of market and credit risk factors is taken. Some risk factors may influence both market and credit risk. Interest rates, for example, are market prices determining the values of various fixed income instruments, but they also have an influence on default probabilities, and they are in turn influenced by idiosyncratic properties of individual obligors.

The used scenarios have been created with *Algo Scenario Engine (ASE)*, using the models described in Table 5.2.

Table 5.2 Risk Factors and Assigned Simulation Models	
Risk Factors	Simulation Model
Spot foreign exchange rates	Geometric Brownian motion
Interest rate treasury curve	Mean-reverting (Black–Karasinski)
Credit drivers	Brownian motion

The Monte Carlo scenarios have been generated using the *standard random sampling (SRS)* method, based on a pseudo-random generator to generate its sequel of random variables. To ensure that the scenarios are reproducible, a seed was specified to the number generator. *SRS* randomly draws its number from a uniform distribution, until reaching the desired number of scenarios, each scenario having the desired number of trigger times. A transformation is applied to convert the uniform random distribution into a normal one. For each scenario and each time step, a random normal vector is created.

Scenarios have been generated on market risk factors like treasury interest rates, Exchange spot rates and credit risk factors like the systemic credit drivers. In this portfolio, 50,000 scenarios been generated on each market and credit state drivers for two time steps of 89 and 365 days. Thereafter, stress test has been conducted on the portfolio with the market risk scenarios and exposures have been generated on the portfolio returns conditional to the future market risk factors and by using the discounted cash flow technique for bond pricing. The loss measure is taken at 99.95% of the loss distribution for the portfolio across time steps and scenarios for all the instruments in the portfolio.

Credit risk is measured with respect to credit drivers' indexes, which are the systemic macroeconomic factors that have an impact on the credit risk of the bond instruments for each counterparty. The region–sector assignment allows the corresponding region–sector index to be associated with each name as a credit driver for the purpose of CWI estimation.

In the portfolio of fixed rate corporate bonds, the positions of bonds will simultaneously depend on market and credit risk factors. This will result into incorrect estimation of true portfolio risk measured on an individual basis. One is interested to assess the impact of the correlation of the market and credit risk on the joint estimation of the integrated risk on the portfolio. The next steps have been followed in order to calculate the integrated market and credit risk:

1. Generate market scenarios—risk factors include primary market risk factors. Scenarios explicitly define the joint evolution of all the relevant market risk factors over the analysis period and are created in this step using historical data and a model. This step contributes to the estimation of exposure at default (the amount to be lost in the event of default, before recoveries are taken into account).
2. Evaluate net credit exposures—the amounts that will be lost in the event of a default or credit migration are computed under each scenario. This step includes many facets of loss given default (loss due to default of counterparty, depending on the amount recovered and the timing of recovery, determined by the seniority of the claim, collateral, etc.).
3. Generate credit scenarios—the fundamental credit model relates obligor to both systematic and idiosyncratic factors. Default probabilities vary as a result of changing economic conditions—and these changes can be viewed as the drivers of default correlations. Systematic factors are simulated consistently with each market scenario. We have used approximately 10,000 market and credit scenarios to capture tail risks appropriately. This step takes care of macroscopic correlation among credit ratings of counterparties and allows integration of market and credit risks. The creditworthiness is obtained based on the credit scenarios upon which an obligor's default probabilities are conditioned and on the scenario path up to each point in time. Correlations among obligor defaults and transitions are determined by their individual relationships to the set of common risk factors, as per Eq. (5.3).
4. Count idiosyncratic risks—conditional on a particular scenario, idiosyncratic risk for each obligor is independent of that for other obligors and is used to assess the actual credit state of the issuer under each credit risk scenario.
5. Aggregate loss distributions and measure risk.

5.5 RESULTS AND SUMMARY

The total exposure of all the instruments for this portfolio is approximately 635 billion USD. As mentioned before, Monte Carlo scenarios are automatically generated by shifting underlying risk factors in a way that is consistent with volatilities and correlations. For this kind

Table 5.3 Calculated VaR for Market, Credit, and Integrated Market and Credit Risk		
VaR Horizon	89 Days	365 Days
Market VaR at 99.95	104,312,600,551.86	140,129,317,599.76
Credit VaR at 99.95	137,898,249.00	198,598,398.50
Integrated VaR at 99.95	309,373,064.50	310,046,865.20

of *VaR*, instruments are accurately revalued under each scenario. The results for the market, credit, and integrated market and credit risk are presented in Table 5.3.

We observe that the market VaR is approximately 104 billion USD for 89 days, which is approximately 16% of the total portfolio value and 140 billion for 365 days, which is approximately 22% of the total portfolio value.

There is a clear indication of adverse market conditions which may have a heavy impact on the portfolio losses due to future movements of market risk factors like the treasury interest rates and the foreign exchange spot rates. We also observe that the market VaR increases for 365 days, which is because of the fact that longer maturity comes with higher level of uncertainty reinforcing the fact that the overall market conditions do not seems to be favorable for the next 1 year

One can observe that the credit VaR at 99.95% for 89 days is approximately 138 million and 199 million for 365 days. The integrated market and credit VaR at 99.95 percentile of the loss distribution is 309 million for 89 days and is 310 million for 365 days, which is approximately very close to that for 89 days.

The above difference between the market and credit risk and the integrated market and credit risk is clear evidence that the market and credit risk factors are heavily correlated.

The market risk VaR is well supported by the dynamic macroeconomic environment wherein the US Fed is revising rates very frequently to deal with inflation, exchange rate instability, and to regain investor confidence once again due to the credit crunch, which started from the month of August 2007 and has had a very marked impact on the global economy since then.

The low estimate of the credit risk might be due to the portfolio consisting of counterparties with very low stochastic probability to default, in spite of having all sorts of credit ratings and due to neglecting the migration risk. On the other side, the input for the credit sensitivity used for this analysis is based on data collected in the past 6 years, which, due to very low rates of defaulting, did not allow a proper modeling of neither the transition probability between the credit states of the counterparties nor the correlation matrix between the different credit risk factors. Now, with so many companies defaulting, there is a clear opportunity to better establish the correlation matrices related to credit risk which, based on low amounts of previously available data, were not very well calculated in order to make possible a forecast of the actual credit crunch.

REFERENCES

Aziz, J., Charupat, N., 1998. Calculating credit exposure and credit loss: a case study. Algo Res. Q. 1 (1), 31–46.

Gordy, M., 1998. A comparative anatomy of credit risk models, Federal Reserve Board, Finance and Economics Discussion Series, Series 1998-47, Board of Governors of the Federal Reserve System (US), pp. 197–232.

Kane, S., Krupp V. and Macki, J., 2001. Monte Carlo Simulation in the Integrated Market and Credit Portfolio Model [Study Group Report] (http://www.maths-in-industry.org/miis/172/), Accessed on December 5, 2011.

CHAPTER 6

A Quantum Mechanics Model May Explain the Infringement of Some Financial Rules in Spite of Stiff Supervision

Radu Chişleag

Bucharest Polytechnic University, Faculty of Applied Sciences, Romania

6.1 INTRODUCTION

Before using the tunnel effect to explain the infringement, the chapter will first introduce the model to interdisciplinary readers not having a modern physics education. We shall have, on this occasion, to introduce the simplest physical model of a one-dimensional potential step is presented first and secondly the model of a potential barrier facing the motion of a quantum particle. To this end, we will follow standard approach of quantum mechanics. Subsequently, the physical characteristics will be replaced by the author with human and social corresponding ones, and the physical conclusions will be interpreted in social and human terms, with emphasis on reference to financial relationships (Chişleag, 2003).

6.1.1 The Potential Step

Let us compare the quantum and the classical predictions in the physical one-dimensional case.

The simplest system is the motion of a particle, of mass m, in a conservative field of potential $V(x)$, of the form illustrated in

Figure 6.1 by the dotted curve. Since the force $F(x)$ deriving from the potential is

$$F(x) = -\frac{\partial V}{\partial x} \tag{6.1}$$

the particle moves freely, except in the neighborhood of the origin, 0, where it is subjected to a force toward the left, opposing its motion. If the particle has the total energy E_0 and the kinetic energy T, then

$$E_0 = T(x) + V(x) \tag{6.2}$$

There are two cases to be considered and we look at them both classically first.

6.2 CLASSICAL APPROACH

Case (i): $E_0 > V$ (classical)

Particles coming from the left approach the potential step with energy higher than the potential step height. The kinetic energy T_0 and the linear momentum P_0 are given by

$$T_0 = E_0 = \frac{p_0^2}{2m} \tag{6.3a}$$

As the particles have sufficient energy to jump on the step, in spite of being slowed up by the force, they move through the region of the potential step, in spite of some of the initial kinetic energy having been converted into potential energy. There is total transmission. The particles emerge to the right with smaller kinetic energy, T_1, and momentum, p_1, where

$$T_1 = \frac{p_1^2}{2m} = E_0 - V \tag{6.3b}$$

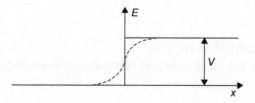

Figure 6.1 The energy diagram of a potential step. The dotted curve gives the actual situation. The solid line is an idealized situation, for which calculations are easier.

Case (ii): $E_0 < V$ (classical)

Particles coming from the left approaching the potential step with an energy smaller than the potential step height are stopped by the potential step at the point x', where

$$V(x') = E_0, \quad [T(x')] = 0 \tag{6.4}$$

Their motion is then exactly reversed under the action of the force. In this case, there is, therefore, total reflection of the beam of particles. The qualitative features of the classical motion are unchanged if the gradual potential step is replaced by a sudden potential step, as illustrated by the solid line in Figure 6.1. This is a simpler system to discuss quantum mechanically.

The Physics–economics correspondence could be, for our purpose:

- Particle = individual; payer, broker, hedger.
- Energy = energy, capability, potential, maximum credit permitted the individual to engage in a process.
- Kinetic energy = capability of the individual to act at a given stage of the process.
- Potential energy = energy used by the individual to jump in the other region (on the potential step).
- Mass = the inertia of the individual.

The conclusions of this classical approach are that:

Case (i): The broker or the hedger acts freely, by observing the rules and disposing of enough funding.

Case (ii): The broker or the hedger has to stop his action at a limit, resulting from the rules and his permitted credit.

That means that an individual who is not permitted to engage more funds at that stock market operation will not play; if he tries to jump on the step, infringing the rules, operating outside the available credit, he will be rejected. This classical physics approach could not explain breaking the rules, the action of the individual being rejected at the step, when not having the permission to act.

6.3 QUANTUM MECHANICAL APPROACH: SCHRÖDINGER REPRESENTATION

The quantum mechanical motion is determined by the Schrödinger energy *eigenvalue* equation

$$\left(\frac{\hat{p}^2}{2m} + V(\hat{x})\right)u_E = E_0 u_E \tag{6.5}$$

where

$$\begin{aligned} V(x) &= 0, & x < 0 \\ V(x) &= V, & x > 0 \end{aligned} \tag{6.6}$$

In the Schrödinger representation, i.e.,

$$\left(-\frac{\hbar^2}{2m}\frac{\partial^2}{\partial x^2} + V(x)\right)u_E(x) = E_0 u_E(x) \tag{6.7}$$

where h bar is a universal constant equal to $1.05 \cdot 10 - 34$ J s.

We have the general boundary condition that $u_E(x)$ must be finite everywhere. We also have to consider the discontinuity at $x = 0$. Since the sudden change in $V(x)$ is finite and $u(0)$ is finite, Eq. (6.7) asserts that $u''(0)$ (the second derivative) is finite. This implies that both $u(x)$ and $u'(x)$ are continuous at $x = 0$. We must now distinguish the two cases.

Case (i): $E_0 > V$ (quantal)

Particles coming from the left approach the potential step with energy smaller than the potential step height.

Define

$$k_0^2 = \frac{2mE_0}{\hbar^2} \tag{6.8}$$

and

$$k_1^2 = \frac{2m(E_0 - V)}{\hbar^2} \tag{6.9}$$

Then Eq. (6.7) becomes, in the left and right regions, respectively

$$\left(\frac{\partial^2}{\partial x^2} + k_0^2 \right) u_L(x) = 0, \quad x < 0 \tag{6.10}$$

$$\left(\frac{\partial^2}{\partial x^2} + k_1^2 \right) u_R(x) = 0, \quad x > 0 \tag{6.11}$$

the subscripts L and R denoting the solutions to left and to the right of the origin, respectively.

The solutions are linear combinations of

$$u_L(x) = e^{\pm ik_0 x} \tag{6.12}$$
$$u_R(x) = e^{\pm ik_1 x} \tag{6.13}$$

These are de Broglie waves corresponding to the linear momenta p_0 and p_1 of the classical problem. We are interested in the situation in which a particle approaches from the left and may then be either transmitted or reflected. We thus look for a solution of the form

$$u_L(x) = \underset{\text{(incident)}}{e^{ik_0 x}} + \underset{\text{(reflected)}}{A\,e^{-ik_0 x}} \tag{6.14}$$

$$u_R(x) = \underset{\text{(transmitted)}}{B\,e^{ik_1 x}} \tag{6.15}$$

We have arbitrarily normalized the coefficient of the incident wave to unity and have neglected the reflected wave in the right region, because no particle returns from infinity.

We want to find the possible energy values, E_0, of the system, and the reflected and transmitted intensities, determined through A^2 and $|B|^2$, respectively.

The continuity conditions at $x = 0$ are

$$1 + A = B \quad \text{(u continous)} \tag{6.16}$$
$$k_0(1 - A) = k_1 B \quad \text{(u' continuous)} \tag{6.17}$$

These equations may be solved for any value of E_0 and imply

$$A = \frac{k_0 - k_1}{k_0 + k_1}, \quad B = \frac{2k_0}{k_0 + k_1} \tag{6.18}$$

Thus, provided $E_0 > V$, the system can have any energy, as in the classical theory.

However, the possible motion differs from the classical in an essential way. The relative probability of finding the particle at a point x, before the potential step, for $x < 0$, is

$$P_\psi(x) = \left| e^{ik_0x} + A\, e^{-ik_0x} \right|^2 = 1 + |A|^2 + 2A \cos 2k_0x \qquad (6.19)$$

The final, oscillating, term is not of much physical interest when discussing the average behavior and can be removed by averaging over a region of length large compared with $2\pi/k_0$. The other two terms come directly from the incident and reflected beams, and may be interpreted as the relative intensities of these beams. By the same token, $|B^2|$ is the relative intensity of the transmitted beam. The important new qualitative feature of the quantum theory is that, since

$$|A|^2 \neq 0 \qquad (6.20)$$

there is a nonvanishing reflected beam.

This corresponds to the situation when some individuals, even having been granted the right to act, are rejected at the step, which happen quite often, situation which may not be described by a classical physics model.

Two limiting cases are of particular interest.

The condition for quantum mechanics to be necessary is

(typical length) × (typical momentum) $\leq \hbar$.

The typical length is the distance through which the potential is changing; the typical momentum is that of the incoming beam. Thus the classical limit is obtained for large momenta or

$$E_0 \gg V \qquad (6.21)$$

In this case, by Eqs (6.8) and (6.9),

$$k_0 \cong k_1$$

Hence, by Eq. (6.18),

$$A \cong 0, \quad B \cong 1 \qquad (6.22)$$

which is the correct classical limit of total transmission.

The individuals having a high energy (more available resources than the step) may pass on the step, with a large probability. An extreme quantum limit is

$$E_0 \gg |V| \tag{6.23}$$

This interesting case means V is large in magnitude, but negative, so that we have a sudden large potential drop (step down!), through which classical particles pass with greatly increased momentum. Classically, for instance the individuals could easily go down and continue playing; they do not return when there is no restriction to play and the step down corresponds to something very attractive. However, quantum mechanically, by Eqs (6.8) and (6.9), we now have

$$k_0 \ll k_1 \tag{6.24}$$

Thus, by Eq. (6.18),

$$A \cong 1, \quad B \cong 0 \tag{6.25}$$

implying total reflection—the exact opposite of the classical prediction. This essential quantum effect may be observed in nuclear physics when a low-energy incident neutron, say, is reflected by the sudden onset of the highly attractive potential, as it approaches the surface of a nucleus.

A similar situation is that of small players who are free to play, but when attracted by huge possibility of gain feel almost completely unable to play losing the chance of gaining. Only important players $k_1 = \sim k_0$ are able to make profit from such opportunities.

Nevertheless, the small players are rather successful when attracted by small gains: $A < 1$; $B < 1$. Because there is a dynamic component of the stock market too, the oscillating term in Eq. (6.19), $2A \cos 2k_0 x$ may become of essential gaining factor, usually accessible to professional players, only. Space does not allow further investigation of this here.

Case (ii): $E_0 < V$ (quantal)

Particles are coming from the left approach the potential step with energy smaller than the potential step height.

The equation for $u_L(x)$, $(x < 0)$, is exactly as before. For $u_R(x)$, we now define

$$K^2 = \frac{2m}{\hbar^2}(V - E_0) \tag{6.26}$$

and then

$$\left(\frac{\partial^2}{\partial x^2} - K^2\right)u_R(x) = 0, \quad x > 0 \tag{6.27}$$

Again, to represent a state with unit intensity incident beam from the left, we have a solution of the form

$$u_L(x) = e^{ik_0 x} + A\,e^{-ik_0 x} \tag{6.28}$$

$$u_R(x) = C\,e^{-Kx} + D\,e^{+Kx} \tag{6.29}$$

To satisfy the condition that u is normalizable, u_R must go to zero at infinity, so we must have $D = 0$.

The continuity conditions at $x = 0$ are

$$1 + A = C \tag{6.30}$$

$$ik_0(1 - A) = -KC \tag{6.31}$$

which can be solved to give

$$A = \frac{k_0 - iK}{k_0 + iK}, \quad B = \frac{2k_0}{k_0 + iK} \tag{6.32}$$

for any value of E_0. Thus again there is no restriction on the possible energy values.

Further

$$|A|^2 = 1 \tag{6.33}$$

so that for any energy, in this range, we have total reflection, as in the classical case.

The broker could not act (risk) on behalf of the bank in that operation. However, the relative probability for finding the particle in the classically forbidden region, $x > 0$, is

$$P_{u_R}(x) = \left|C\,e^{-Kx}\right|^2 = \frac{4k_0^2}{k_0^2 + K^2} \times e^{-2Kx} \tag{6.34}$$

This probability is appreciable near to the step edge and falls off exponentially to a negligible value over a distance that is large compared with $1/k$.

The effect is particularly important if we consider a wave impinging on a barrier of finite thickness (Figure 6.2).

This effect will give an appreciable probability for finding the particle at the opposite edge of the potential barrier, and the particle will then propagate to the right as a free particle.

From Eq. (6.34), the relative probability of finding the particle at $x = b$ with respect to $x = 0$ is

$$T = \exp[-2Kb] = \exp\left[-2\left(\frac{2m}{\hbar^2}\right)^{1/2}(V-E_0)^{1/2}b\right] \qquad (6.35)$$

This is an approximate expression, valid for large b, for the probability that a particle, of energy E_0, penetrates a barrier of height V and width b. The solution inside a barrier of finite width is really a mixture of rising and falling exponential ($\exp[\pm kx]$), and for narrow barriers there is a considerable contribution of the rising exponential. For some defined values of energy, the transmission reaches unity (resonances occur).

This quantum possibility of penetrating potential barriers, which will certainly stop classical particles, is the basis in physics of the explanation of the radioactive decay of a nucleus. It is known as tunnel effect.

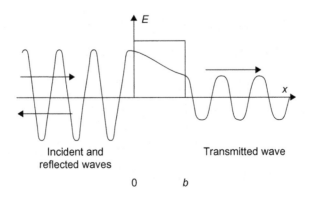

Figure 6.2 The energy diagram of a finite potential barrier, showing the incident and transmitted waves.

The transmission by tunnel effect (6.35) is increasing when the energy gets closer to the barrier potential height, when the width of the barrier gets thinner, more, when the mass (inertia, single broker implied) is smaller. The tunnel effect may explain the infringing of financial rules by some players, accompanied by huge gains (usually remaining anonymous) or by huge losses (widely publicized in the media!).

Eventually, large relative gains of some small players happen. The width of the potential barrier is also subject of alteration, by relaxation or strengthening of the audit. Usually, the internal audit implies a large barrier (Figure 6.3).

The successive gains of a player may induce the relaxation of the internal check of the bank—that means of the width of the barrier, the system passing from the situation described by Figure 6.3 to the situation described by Figure 6.4, which largely increases the opportunity for the authorized player to assume higher risks, to gain huge amounts

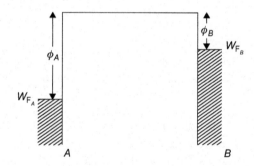

Figure 6.3 Wide barrier of potential between two areas at different levels of potential.

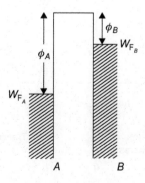

Figure 6.4 Thin barrier of potential between two areas at different levels of potential.

of money but also to lose huge sums, eventually outside the coverage of the supporting financial body.

Following a huge loss, the check is improved again, the barrier's width is increased very much (goes back to the situation like in Figure 6.3) and the probability of losing huge sums is decreased significantly (of course, of very high relative gains, too). Only individuals with higher personal potential (results), assuming high risks, would be able to cross the barrier of the internal audit, but with a much smaller probability to do this, until a new relaxation of the rules or the financial support would happen. This physical model may be extended from rectangular barriers to barriers having different profiles. The physical conclusions are close to the present ones. The investigation of the role of the oscillating term and of possible resonances would be the topics of further research. A special class of barriers, of those having an axis of symmetry (symmetrical potential), is very important because the solutions to the Schrödinger equation split themselves in two distinct classes: symmetrical and antisymmetrical solutions.

As is shown in quantum statistics, the symmetrical solutions are describing the behavior of bosons (system particles able to occur in any number in a given state) and fermions (system particles that can occur in different states only). This distinction between bosons and fermions would be important for players, eventually corresponding to those cooperating or respectively competing. This quantum physics model, introduced first to explain some aspects of international relations, mainly during the Iron Curtain period, subject to a good quantification of the socio-human "energy", might be useful too in many social and human fields of study.

The rationale behind this is connected with the Heisenberg Principle of uncertainty, which may be valid in social and human behaviors: the actual behavior of an individual alone in a given action is much more difficult to forecast than the average behavior of an individual belonging to a social group, in repeated actions.

REFERENCE

Chişleag, R., 2003. A quantum mechanical model to explain the infringement of barriers impeding international relationships. Cd-Rom Proceedings of the Eci & E4 Conference "Enhancement of the Global Perspectives for Engineering Students", Tomar, Portugal, April 6 – 11, Carl McHargue University of Tennessee, USA and Eleanor Baum, Cooper Union, USA, Eds, ECI Symposium Series, Volume P03 (2003). < http://dc.engconfintl.org/enhancement/19 > Last (accessed 02.10.2012.)

CHAPTER 7

The Potential of Econophysics for the Study of Economic Processes

Gheorghe Săvoiu[1] and Constantin Andronache[2]

[1]University of Piteşti, Faculty of Economics Romania; [2]Boston College, Chestnut Hill, MA, USA

7.1 INTRODUCTION

Specific economic theories are constructed from rationality which is the same thing as maximizing expected utility by applying the basic postulates to various economic situations. The new science called econophysics by Rosario Mantegna and H. Eugene Stanley, at the second Statphys-Kolkata conference in 1995, follows the path of such mergers as astrophysics, biophysics, geophysics, etc. and is based on the observation of similarities between economic systems and concepts and those from physics, using the methods from statistical physics (Chakrabarti, 2005). There is nothing new in the close relationship between physics and economics.

A lot of the great economists did their original training and took their models in and from physics, and the influence of physics is clearly evident in many of economic theory's models. But it is well known too that physics had the more dominating effect on the development of formal economic theory. Its molecular statistical theory of thermal phenomena furthered the cause of mechanical atomism and population-based thinking.

The empirical and theoretical impact of the relativistic and quantum frameworks of physics cemented new, deeper boundaries. Relativity captured space, time, and gravitational interaction; quantum mechanics, matter and the rest of forces (*Stanford Encyclopedia of Philosophy*). The adequate use of concepts and methods of demography thinking in economics and the role of the environment have been developed to study a new kind of complex dynamical system like the economic or the financial one.

The relation that has transpired between economics and physics over the past two decades seems very likely to be a model for the future, and the purpose of this chapter is to identify the potential of econophysics through its new domains and results.

7.2 WHAT IS THE ADEQUATE MEANING FOR CONTEMPORARY ECONOPHYSICS?

The field of research known as econophysics has alternative names such as financial physics and statistical finance, this arising initially from its being a new development of two different disciplines like finance and physics. But in the last few years, more and more work has been done outside the field of finance. Rosario Mantegna and Eugene H. Stanley have proposed the first definition of econophysics as a multidisciplinary field or "the activities of physicists who are working on economics problems to test a variety of new conceptual approaches deriving from the physical sciences" (Mantegna and Stanley, 2000).

From the classical physics side, econophysics is mainly considered to be restricted to the principles of statistical mechanics, which is the application of probability theory to large numbers of physical objects which are related to each other in a certain way, while from the economics side, macroscopic properties are viewed as the result of interactions at the microscopic level of the constituents, and econophysics becomes for economists that have always encouraged the application of quantitative and formal methods "the investigation of economic problems by physicists" (Roehner, 2005). But somehow econophysics is more interesting viewed by its practitioners, as a revolutionary reaction to standard economic theory that threatens to enforce a paradigm shift in thinking about economic systems and phenomena. Yakovenko's (2009) relevant definition considers that econophysics is an "interdisciplinary research field applying methods of statistical physics to problems in economics and finance."

Econophysicists, making use of the statistical physics analogy, rather than a thermodynamic one, adopt a constructive-theory-type approach. And thus their pioneering models are microscopically realistic, and where the parameters hold some physical meaning they are derived from the data and are physically well founded by providing basic mechanisms for the phenomena. Modern econophysics, a very rapidly developing area, proposes the application of methods from statistical physics, the physics of complex systems, and science of networks to macro/microeconomic modeling, financial market analysis, and social problems (from demographical to cultural, from real economical convergence to social cohesion problems, etc.). Indeed, an extended contemporary definition of econophysics define this new science as an interdisciplinary research field, applying theories and methods originally developed by physicists in order to solve problems in economics, such as those including uncertainty or stochastic elements and nonlinear dynamics. Its application to the study of financial markets has also been termed statistical finance referring to its roots in statistical physics.

The contemporary econophysics perspective applies to economic phenomena, various models and concepts associated with the physics of complex systems (e.g., statistical mechanics, condensed matter theory, self-organized criticality, and microsimulation). Econophysics will become more and more a translation of statistical physics into real individuals and economic reality. Where the curious statistical properties of the data are fairly well known among economists, but still remain a puzzle for economic theory, that will be the right place where physicists must come on stage with their method and techniques... For econophysicists, the universality of the statistical properties is their starting point, and for this new and stable vision of the truth, the most recently awarded Nobel Prizes, in the last decade, recognize outstanding original contributions that use physical methods to develop a better understanding of socioeconomic problems.

7.3 A BRIEF HISTORICAL BACKGROUND OF ECONOPHYSICS

For the new econophysics, its first applications have been almost invariably in the financial markets, which certainly comprised many different actors, engaged in relatively frequent interactions. Statistical mechanics or physics was developed in the second half of the nineteenth century by James Clerk Maxwell, Ludwig Boltzmann, and Josiah Willard Gibbs. These physicists developed mathematical methods for describing the atoms statistical properties: the probability distribution of velocities of molecules in a gas (the Maxwell–Boltzmann distribution), and the general probability distribution of states with different energies (the Boltzmann–Gibbs distribution).

The role of physics models as the foundations for the standard neoclassical model that current econophysicists seek to displace is much older than two centuries:

- N.-F. Canard wrote, in 1801, that supply and demand were ontologically like contradicting physical forces.
- Léon Walras was deeply influenced by the physicist Louis Poinsot in his formulation of this central concept of general equilibrium theory in economics.
- Irving Fisher, father of American mathematical economics in its neoclassical form, was a student of Willard Gibbs, the father of statistical mechanics.

The interest of physicists in financial and economic systems has roots that date back to 1936, when Majorana wrote a pioneering paper, published in 1942, entitled *Il valore delle leggi statistiche nella fisica e nelle scienze sociali*, on the essential analogy between statistical laws in physics and social sciences. Many years later, a statistical physicist, Elliott Montroll, coauthored with W.W. Badger in 1974 the book *Introduction to Quantitative Aspects of Social Phenomena*.

The application of concepts as power-law distributions, correlations, scaling, unpredictable time series, and random processes to financial markets was possible during the past two or three decades; physicists have achieved important results in statistical mechanics, nonlinear dynamics, and disordered systems, and also due to other significant statistical investigations and mathematical formalizations. First of all, more than one hundred years ago, an Italian economist and statistician,

Vilfredo Pareto, who earned first a degree in mathematical sciences and a doctorate in engineering, investigated the statistical character of the wealth of individuals in a stable economy, by modeling them using a special distribution ($Y = x - V$), where Y is the number of people having income x or greater than x and v is an exponent that Pareto estimated to be 1.5.

Secondly, the mathematical formalization of a random walk was published by Louis Bachelier in his doctoral thesis entitled *Théorie de la spéculation*, at the Academy of Paris, on March 29, 1900, in which Bachelier determined the probability of price changes. The first description of a random walk made by a physicist was performed in 1905 by Albert Einstein and the mathematics of the random walk was made more rigorous by Norbert Wiener. Bachelier's original proposal of Gaussian distributed price changes was soon replaced by a lot of alternative models, from which the most appreciated was a geometric Brownian motion, where the differences of the logarithms of prices are distributed in a Gaussian manner (Mantegna and Stanley, 2000).

Since the 1970s, a series of significant changes has taken place in the world of finance that finally will be born the new scientific field of econophysics. One key year was 1973, when currencies began to be traded in financial markets, and the first paper was published that presented a rational option-pricing formula (Black and Scholes, 1973).

A second revolution began in the 1980s, when electronic trading was adapted to the foreign exchange market and the result has become a huge amount of electronically stored financial data readily available. Since the same 1980s, it has been recognized in the physical sciences that unpredictable time series and stochastic processes are not synonymous. Chaos theory has shown that unpredictable time series can arise from deterministic nonlinear economic systems and theoretical and empirical studies have investigated whether the time evolution of asset prices in financial markets might indeed be due to underlying nonlinear deterministic dynamics of a relative limited number of variables.

Since the 1990s, a growing number of physicists have attempted to analyze and model financial markets and, more generally, economic systems, and new interdisciplinary journals have been published, new conferences have been organized, and a lot of new potentially scientific fields, areas, themes, and applications have been identified. The researches of econophysics deal with the distributions of returns in financial markets, the time correlation of a financial series, the analogies and differences between price dynamics in a financial market and physical processes as turbulence or ecological systems, the distribution of economic stocks and growth rate variations, the distribution of firm sizes and growth rates, the distribution of city sizes, the distribution of scientific discoveries, the presence of a higher-order correlation in price changes motivated by the reconsideration of some beliefs, the distribution of income and wealth, the studies of the income distribution of firms, and studies of the statistical properties of their growth rates.

The statistical properties of the economic performances of complex organizations such as universities, regions, or countries have also been investigated in econophysics. The new real characteristics of the econophysics on a medium and long term will be a result of its new research, like rural–urban migration or growth of cities. The real criticism of econophysics is the absence of age variable, because models of econophysics consider immortal agents who live forever, like atoms, in spite of evolution of income and wealth as functions of age, that are studied in economics using the so-called overlapping-generations models. Even though with the time both physics and economics became more formal and rigid in their specializations, and the social origin of statistical physics was forgotten, the future is perhaps a common one.

On the computer, econophysicists have established a niche of their own, by making models much simpler than most economists now choose to consider even using possible connection between financial or economical terms and *critical points* in statistical mechanics, where the response of a physical system to a small external perturbation becomes infinite because all the subparts of the system respond cooperatively, or the concept of "noise", in spite of the fact that some economists even claim that it is an insult to the intelligence of the market to invoke the presence of a noise term. Many different methods and techniques from physics and the other sciences have been explored by econophysicists, sometimes frantically, including chaos theory, neural networks, and pattern recognition.

Econophysics also means a scientific approach to quantitative economy using ideas, models, conceptual, and computational methods of statistical physics. In recent years, many physical theories like theory of turbulence, scaling, random matrix theory, or renormalization group were successfully applied to economy giving a boost to modern computational techniques of data analysis, risk management, artificial markets, and macroeconomy (Burda et al., 2003). And thus econophysics became a regular discipline covering a large spectrum of problems of modern economy.

But even today in this new era of econophysics there still remains a negative impact of physics with economics for which both physicists and economists are in part responsible, because of the failure of economists to deal properly with certain empirical regularities and a lot of economists still have a mind set which is unusually closed, or the fact that many physicists cannot understand even the simplest supply-and-demand model, or the fact that physicists and economists belong to the distinct categories of physical or natural (hard) science and social (soft) science, etc. Science or financial markets are only a very small part of economic theory and some physicists naively believe and search for universal empirical regularities in economics that probably do not exist and seem to have been reluctant to work in areas where data sets are short and unreliable, but this characterizes a great deal of data in the social sciences and economics.

7.4 METHODS OF ECONOPHYSICS

Contemporary econophysics involves in effect physicists doing economics with theories from physics, which raises the question of how the two disciplines relate to each other, and accounting for interest rates and fluctuations of stock market prices; these theories draw analogies to earthquakes, turbulence, sand piles, fractals, radioactivity, energy states in nuclei, and the composition of elementary particles (Bouchaud, 2002). Today it becomes possible for methods and concepts of statistical physics to have real influence in economic thought, but it is also possible that economical methods and concepts can influence physics thought too. The method of econophysics defines its main goal in applying method of statistical physics and other mathematical methods used in physics to economic data and economic processes. Why can the methods and techniques from statistical physics can be successfully applied to social, economical, and financial problems? It could be the result of the great experience of physicists in working with experimental data gives them a unique advantage to uncover quantitative laws in the statistical data available in social sciences, economics, and finance? Is indeed econophysics bringing new insights and new perspectives, which are likely to revolutionize the old social sciences and classical economics?

The study of dynamic systems is mostly based on expressing them in terms of (partial) differential equations which are further solved by analytic methods (or numerically). But this is somehow against our intuitions: we never meet in our life density distributions of our friends, cars, utility functions, etc. We have converted integers into real numbers by averaging over certain areas. This can be done either by averaging over large enough volumes or over long period of times. Statistical physics is a framework that allows systems consisting of many heterogeneous particles to be rigorously analyzed. Inside econophysics, these techniques are applied to economic particles, namely investors, traders, and consumers. Markets are then viewed as (macroscopic) complex systems with an internal (microscopic) structure consisting of many of these particles interacting so as to generate the systemic properties (the microstructural components being reactive in this case, as mentioned already, thus resulting in an adaptive complex system). When the first physicists tried to analyze financial markets applying the method of statistical physics, they did not view these markets as particularly fine examples of complex systems, as cases of complexity in action. Some of them have even believed they are discovering laws or some stability evidence in the form of the scaling laws that Pareto first investigated (but that has been found in a much wider variety of economic observables). In truth, the stability evidence discovered or the empirical distribution is not a stable or definitive one (a conclusive one), because all the markets are characterized by nonstationarity, that is a general feature of adaptive complex systems: "the empirical distribution is not fixed once and for all by any law of nature [but] is also subject to change with agents' collective behaviour" (McCauley, 2004a). Theory confirms that characteristics of complex systems involve three necessary conditions:

1. Complex system must contain many subunits (the exact number being left vague).
2. Subunits must be interdependent (at least some of the time).
3. Interactions between the subunits must be nonlinear (at least some of the time).

These properties are said to be emergent when they amount to new complex or systemic structure and an adaptive complex system adds the following condition:

• Individual subunits modify their properties and behavior with respect to a changing environment resulting in the generation of new systemic properties.

Finally, the organizing adaptive complex system adds an important condition also:

• Individual subunits modify their own properties and behavior with respect to the properties and behavior of the unit system they jointly determine, sometimes introducing a network efficiency measure for what are referred to, by physicists, in particular, as complex networks (Latora and Marchiori, 2003, 2004).

In a comparison to classical statistical thought, econophysics has revealed that heterogeneous in reality must be explained with homogeneous in theory. And this is the main role of the method of statistical physics — to unify and simplify economics.

7.5 REVIEW OF MAJOR RESULTS AND NEW DOMAINS

From the perspective of the authors (an economist and a physicist), the two main elements of econophysics for an update review are the results and the new domains in refereed literature. But in fact it is really difficult to do it properly without two significant opinions. Both are from the most important representatives of the American school of econophysics.

(A) The first opinion belongs to Eugene H. Stanley, the well-known father of the new science (physicist to Boston University, Department of Physics), and it was written during a scientific talk about recent applications of correlated randomness to economics for which statistical physics is proving to be particularly useful:

1. Traditional economic theory does not predict outliers, but recent analysis of truly huge quantities of empirical data suggests that statistical physics does not fail for it. (If statistical physics analyzes only a small data set, not more than 10^4 data points, then outliers appear to occur as rare events, but when orders of magnitude

higher than 10^8 data points are analyzed, a responsible option of statistical physics must not ignore them and studying their properties becomes a realistic goal).

2. In classical economics, neither the existence of power laws nor the exact exponents have any accepted theoretical basis, but the method of econophysics does it.

3. Some economic phenomena described by power-law tails have been recognized for over one hundred years, but it becomes a scientific reality due to statistical physics.

4. Nowadays, the concepts of scaling and universality provide the conceptual framework for understanding the geometric problem of percolation frequently used in econophysics (percolation analysis).

5. Since economic systems comprised of a large number of interacting units have the potential of displaying power-law behavior; it is perhaps not unreasonable to examine economic phenomena within the conceptual framework of scaling and universality.

6. The amassing of empirical facts led to the finding of laws in statistical physics, but finding them is only the first or empirical part of econophysics task, and the second or theoretical part generates more difficulties because it means understanding new laws.

7. While the primary function of a market is to provide a venue where buyers and sellers can transact, the more the buyers and sellers at any time, the more efficient the market is in matching buyers and sellers, so a desirable feature of a competitive market becomes liquidity (defined as ability to transact quickly or to buy at a low price and sell at a high price, according to the prevalent market demand, liquidity that is the main compensation to market makers for the risk they incur). Quantifying the fluctuations of the bid−ask spread, that reflects the underlying liquidity for a particular stock, offers a way of understanding the dynamics of market liquidity.

8. One supplementary reason the economics is of interest to statistical physicists is the system made up of many subunits, like Ising's econophysics model in which subunits are called spins, which is merely to say buyers and sellers, from the classical Ernst Ising contribution to the Theory of Ferromagnetism (Ising, 1925; Palmer, 2007). During any unit of time, these subunits of the economy may be either positive or negative as regards perceived market opportunities for the people that interact with each other, and this fact often produces what economists call the herd effect. The orientation of

whether we buy or sell is influenced not only by our neighbors but also by news (bad news, means to sell). So the state of any subunit is a function of the states of all the other subunits and of a field parameter. On a qualitative level, economists often describe a price change as a hyperbolic-tangent-like function of the demand, which is not quantified, so one of the most important things econophysics had to do was quantify demand. And econophysics did this by analyzing huge databases comprising every stock bought or sold, which gives not only the selling price and buying price, but also the asking price and the offer price.

9. The cross-correlation is another important problem that econophysics has been studying, and that means how the fluctuations of one stock price correlate with those of another.

10. The first model for econophysics was unifying the power laws (large movements in stock market activity arise from the trades of the large participants). Starting from an empirical characterization of the size distribution of large market participants (mutual funds), econophysics show that their trading behavior when performed in an optimal way generates power laws observed in financial data.

11. No one can predict future trends, but approximate inequalities are sometimes predictable. Econophysics, where physicists collaborate with economists and the result is more probable to be useful and responsible, has benefited from collaborations with top-quality energetic economists.

12. Econophysics learns from the work of the masters in science. Econophysicists have always reverenced two outstanding personalities: Vilfredo Pareto, the father of both scaling and universality, and Omori, who discovered relations among aftershocks to an earthquake.

13. Econophysics realizes its contribution of most utility in economics is nothing else but the novelty of thinking about and analyzing data, especially since many methods from mathematical statistics are not focused on handling the strange behavior of nonstationary functions that obey scale invariance, over a limited region of the range of variables (Stanley, 2008).

14. Finally, to find universality in economic phenomena is indeed more possible for methods and concepts of statistical physics (Gabaix et al., 2008; Stanley et al., 2006).

(B) The second is the opinion of Victor Yakovenko (physicist at the University of Maryland, Department of Physics), who identifies the following results:

1. The attention of econophysics was primarily focused on analysis of financial markets and its important achievements define new statistical mechanics of money distribution (starting with fundamental law of the equilibrium statistical mechanics of Boltzmann–Gibbs distribution and finishing with Gamma distribution) (Statphys-Kolkata conference or school and related papers).
2. Econophysics literature has often used, on exchange models, the terms money and wealth interchangeably. While economists emphasize the difference between these two concepts, for the econophysicists wealth is equal to money plus the other property that an agent has. In order to estimate the monetary value of property, econophysics needs to know the price and thus models appear with a conserved commodity, models with stochastic growth of wealth, and so more and more empirical data on money and wealth distributions.
3. Econophysics discovers a lot more empirical data available for the distribution of income from tax agencies and population surveys and so creates new theoretical models of income distribution.
4. If in physics a difference of temperatures allows the creation of a thermal machine, then automatically the difference of money or income temperatures between different countries allows extracting profit in international trade. This process very much resembles what is going on in this new globalized economy where the perpetual trade deficit of the United States is the consequence of the second law of thermodynamics and the difference of temperatures between the United States and the low-temperature countries, such as China.
5. In physics language, the segregation found by Schelling represents a phase transition of the system (similar to interaction energy between two neighboring atoms that depends on whether their magnetic moments point in the same or in the opposite directions), while in economics it becomes transition, and this new concept means that a small amount of one substance dissolves into another up to some limit, but phase separation (segregation) develops for higher concentrations, and thus physicists have decided to be helpful for practical applications of such models (Yakovenko, 2011; Yakovenko and Rosser, 2009).

In the last 5–10 years, the inventory for the new domains of econophysics has been really amazing. Thus, econophysics has dealt with more economic and also noneconomic subjects:

1. A thermodynamic formulation of economics, from N. Georgescu-Roegen (1971) and D.K. Foley (1994 and 1996) to J. Mimkes (2000), using physical terms: capital to energy, GDP per capita to temperature, production function to entropy, etc. (Drăgulescu and Yakovenko, 2001, 2000; Solomon and Richmond, 2001; Aruka, 2001; Smith and Foley, 2008).
2. Zero-intelligence models of limit order markets (Stinchcombe, 2006).
3. Understanding and managing the future evolution of a competitive multiagent population (Smith and Johnson, 2006).
4. The firm's growth and networks (Fujiwara et al., 2003).
5. A review of empirical studies and models of income distributions in society (Richmond, Hutzler, Coelho, and Repetowicz, 2006).
6. Models of wealth distributions—a perspective (Kar Gupta, 2006).
7. The contribution of money transfer models to economics (Wang, Xi, and Ding, 2006).
8. Fluctuations in foreign exchange markets (Aiba and Hatano, 2006).
9. Stock and foreign currency exchange markets (Caldarelli et al., 1997; Ausloos, 2000).
10. Computer simulation of language competition by physicists (Stauffer and Schulze, 2005).
11. Social opinion dynamics (Hegselmann–Krause model; Hegselmann and Krause, 2002; Hegselmann, 2004).
12. Opinion dynamics, minority spreading, and heterogeneous beliefs (Galam, 2005).
13. Persuasion dynamics (Weisbuch et al., 2005).
14. How a hit is born: the emergence of popularity from the dynamics of collective choice (Sinha and Kumar Pan, 2006).
15. Dynamics of crowd disasters (Helbing et al., 2007).
16. Complexities of social networks: a physicist's perspective (Sen, 2006).
17. Emergence of memory in networks of nonlinear units: from neurons to plant cells (Inoue, 2006).
18. Self-organization principles in supply networks and production systems (Helbing et al., 2006).
19. Econophysics of precious stones (Watanabe et al., 2006).

20. Econophysics of interest rates and the role of monetary policy (Cajueiro and Tabak, 2010).
21. Quantum econophysics (Guevara, 2006).
22. Inverse statistics in the foreign exchange market or inverse statistics in econophysics (Andersen; Jensen et al., 2004).
23. Superstatistics in econophysics (Ohtaki and Hasegawa, 2000).
24. Statistical mechanics of money (Dragulescu and Yakovenko, 2000).
25. Global terrorism and underground activities (Galam, 2003).
26. A family-network model for wealth distribution in societies (Coelho et al., 2005).
27. How the rich get richer (Mehta et al., 2005).
28. The emergence of Bologna and its future consequences. Decentralization as cohesion catalyst in guild dominated urban networks (Zimmermann et al., 2001; Zimmermann and Soci, 2003).
29. Basel II for physicists (Scalas, 2005).
30. Advertising in Duopoly Market (Situngkir, 2004).
31. Asymptotic behavior of the daily increment distribution of the Mexican stock market index (Coronel-Brizio and Hernandez-Montoya, 2005).
32. Who is the best connected scientist from the scientific coauthorship networks? (Newman, 2004).
33. Laser welfare: first steps in econodynamic engineering (Willis, 2005).
34. Stability through cycles (De Groot and Franses, 2005, 2006).
35. Agent-based model construction in financial economic system (Situngkir and Surya, 2004).
36. The process of price formation and the skewness of asset returns (Reimann, 2006).
37. Scale-free avalanche dynamics in the stock market (Bartolozzi et al., 2006).
38. A trade-investment model for distribution of wealth (Scafetta et al., 2004).
39. Tobin tax and market depth (Ehrenstein et al., 2005).
40. Complex systems, information technologies, and tourism: a network point of view (Baggio, 2006).
41. Socioeconomic interaction and swings in business confidence indicators (Hohnisch et al., 2005).
42. Statistical laws in the income of Japanese companies (Mizuno et al., 2001).

43. The mechanism of double-exponential growth in hyperinflation and the correlation networks among currencies (Mizuno et al., 2002, 2006).
44. Could the pound and euro be the same currency? (Matsushita et al., 2007).
45. Nonextensive statistical mechanics and economics (Mantegna et al., 1999; Tsallis et al., 2003; Tsallis, 2009).
46. What economists can learn from physics and finance (McCauley, 2004b).
47. Statistical entropy in general equilibrium theory (Liossatos, 2004).
48. The political robustness in Indonesia (Situngkir, 2006).
49. Accelerated growth of networks (Dorogovtsev and Mendes, 2002).
50. Patterns, trends, and predictions in stock market indices and foreign currency exchange rates (Ausloos and Ivanova, 2004).
51. Equilibrium econophysics: a unified formalism for neoclassical economics and equilibrium thermodynamics (Sousa and Domingos, 2006).
52. Predictability of large future changes in a competitive evolving population (Lamper et al., 2008).
53. Correlations between the most developed countries (Miskiewicz and Ausloos, 2005).
54. Are social structures determined by the economy? (Corso et al., 2003).
55. Growth and allocation of resources in economics: the agent-based approach (Scalas et al., 2006).
56. A mathematical framework for probabilistic choice based on information theory and psychophysics (Takashi, 2006).
57. The domino effect for markets (Schulze, 2002).
58. A cluster-based analysis of some macroeconomic indicators in various time windows (Gligor and Ausloos, 2007).
59. Network of econophysicists: a weighted network to investigate the development of econophysics (Fan et al., 2004).
60. The why of the applicability of statistical physics to economics (Guevara, 2007).
61. A brief history of economics: an outsider's account (Chakrabarti, 2006; Chakrabarti et al., 2006).
62. Econophysics: scaling and its breakdown in finance (Mantegna and Stanley, 1997).
63. Consumer behavior and fluctuations in economic activity (Westerhoff, 2005).

64. Trading behavior and excess volatility in toy markets (Marsili and Challet, 2001).
65. Quantum games entropy (Guevara, 2007).
66. Stylized facts of financial markets and market crashes in minority games (Challet et al., 2001).
67. Comment on thermal model for adaptive competition in a market (Challet et al., 2000).
68. Comment on: role of intermittency in urban development (Marsili, Maslov, Zhang, 1998).
69. Econophysics and quantum statistics (Maslov, 2002a).
70. The notions of entropy, hamiltonian, temperature, and thermodynamical limit in probability theory used for solving model problems in Econophysics (Maslov, 2002b).
71. A graph network analysis of gross domestic product (GDP) time correlations (Miśkiewicz and Ausloos, 2006b).
72. A moving-average-minimal-path-length method for EU country clustering according to macroeconomic fluctuations (Gligor and Ausloos, 2008a).
73. Clusters in weighted macroeconomic networks: the EU case (Gligor and Ausloos, 2008b).
74. The application of continuous—time random walks in finance and economics (Scalas, 2006).
75. Microeconomics of the ideal gas like market models (Chakraborti and Chakrabarti, 2009).
76. A phenomenon of concentration—diversification in contemporary globalization (Săvoiu et al., 2012).
77. New metacorrelations (correlations between the intra-market correlations) and interdependencies in the global markets (Kenett et al., 2012).
78. Systemic risk in banking ecosystems during reccesions and crisis (Haldane and May, 2011).
79. Crashes: symptoms, diagnoses, and remedies (Ausloos et al., 2002).
80. Contemporary systemic risk and network dynamics (Abergel et al., 2013).
81. New production networks and geographical economics (Weisbuch and Battiston, 2007).
82. New order-driven markets (Abergel et al., 2011).
83. Rethinking research management: Colombia (Zarama et al., 2007).
84. The cross-correlation distance evolution of the economy globalization (Miśkiewicz and Ausloos, 2006a).

85. Boltzmann−Gibbs distribution of fortune and broken time reversible symmetry in econodynamics (Ao, 2007).
86. Bankruptcy as an exit mechanism for systems with a variable number of components (Gatti et al., 2004).
87. Modeling the information society as a complex system (Baggio, 2006; Olivera et al., 2011).
88. Econophysics and individual choice (Bordley, 2005).
89. Econophysics in the Euromillions lottery (Mostardinha et al., 2006).
90. A natural value unit econophysics as arbiter between finance and economics (Defilla, 2007), etc.

Econophysics will continue to contribute due to its statistical physics method to economics in a variety of different directions, ranging from macroeconomics to market microstructure, and such work will have increasing implications for economic policy making.

7.6 CONCLUSIONS

Within the next few years econophysics will be expected to develop physical method in understanding economical processes, generating new disciplines like demographysics, indexphysics, or physical prognosis, in a so-called economical and physical interaction with other sciences. The new most important domain must be demographysics (international migration), some special area of marketing and management, some distinctive field of index numbers (from poverty index to corruption or globalization index, from Consumer Price Index to Dow−Jones Industrial Average, etc.) and most of all to prognosis or spatial and temporal estimation. The accuracy of physical indices could be improved by statistical physics with its careful thinking in terms of dimensional analysis, combined with better data analysis correlating prices, and other factors, such as wages and pensions, for which the indices are designed.

A new approach will appear in universities too. New basic courses must teach the essential elements of both physics and economics in the new curricula. One area of opportunity will be behavior such as the distribution of wealth or the size of firms, but also new directions like marketing and management. Dimensional and scaling methods have been just a cornerstone in the understanding of complex phenomena like turbulence in fluids, and all the constituents that make fluid flow complex—long-time correlations, nonlinearity, and chaos—are likely to become even greater factors in the economy. The consequences could be a new result in lowering transaction costs and generally making markets more efficient. Understanding the dynamics and statistical mechanics of agency promises to be the key to expanding concepts from economics in physics, too.

REFERENCES

Abergel, F., Chakrabarti, B.K., Chakraborti, A., Mitra, M., 2011. Econophysics of Order-Driven Markets (New Economic Windows). Springer, Milan (Proc. vol. V).

Abergel, F., Chakrabarti, B.K., Chakraborti, A., Ghosh, A., 2013. Econophysics of Systemic Risk and Network Dynamics (New Economic Windows). Springer, Milan (Proc. vol. VI—to be published).

Aiba, Y., Hatano, N., 2006. Fluctuations in foreign exchange markets. In: Chakrabarti, B.K., Chakraborti, A., Chatterjee, A. (Eds.), Econophysics and Sociophysics: Trends and Perspectives. Wiley-VCH, Berlin.

Ao, P., 2007. Boltzmann–Gibbs distribution of fortune and broken time reversible symmetry in econodynamics. Commun. Nonlinear Sci. Numer. Simul. 12 (5), 619–626.

Aruka, Y., 2001. Evolutionary Controversies in Economics. Springer, Tokyo.

Ausloss, M., 2000. Statistical physics in foreign exchange currency and stock markets. Phys. A 285, 48–65.

Ausloos, M., Ivanova, K., 2004. Patterns, trends and predictions in stock market indices and foreign currency exchange rates. In: Wille, L.T. (Ed.), New Directions in Statistical Physics: Econophysics, Bioinformatics, and Pattern Recognition. Springer, Berlin, pp. 93–114.

Ausloos, M., Ivanova, K., Vandewalle, N., 2002. Crashes: symptoms, diagnoses and remedies. In: Takayasu, H. (Ed.), Empirical Science of Financial Fluctuations—The Advent of Econophysics. Springer, Tokyo, pp. 62–76.

Baggio, R., 2006. Complex systems, information technologies, and tourism: a network point of view. J. IT Tourism 8 (1), 15–29.

Bartolozzi, M., Leinweber, D.B., Thomas, A.W., 2006. Scale-free avalanche dynamics in the stock market. Quantitative Finance Papers. <http://arxiv.org/pdf/physics/0601171v2.pdf> Last (accessed 01.10.2012).

Black, F., Scholes, M., 1973. The pricing of options and corporate liabilities. J. Pol. Econ. 81, 637–654.

Bordley, R.F., 2005. Econophysics and individual choice. Phys. A: Stat. Mech. Appl. 354 (1–4), 479–495.

Bouchaud, J., 2002. An introduction to statistical finance. Phys. A: Stat. Mech. Appl. 313 (1–2), 238–251.

Burda, Z., Jurkiewicz, J., Nowak, M.A., 2003. Is econophysics a solid science? Acta Phys. Pol. Ser. B 34 (1), 87–131.

Cajueiro, D.O., Tabak, B.M., 2010. Fluctuation dynamics in US interest rates and the role of monetary policy. Finance Res. Lett. 7 (3), 163–169 (Elsevier).

Caldarelli, G., Marsili, M., Zhang, Y.C., 1997. A prototype model of stock exchange. Europhys. Lett. 40 (5), 479–484.

Chakrabarti, B.K., 2005. Econophys-Kolkata: a short story. In: Chatterjee, A., Yarlagadda, S., Chakrabarti, B.K. (Eds.), Econophysics of Wealth Distributions. Springer, Milan, pp. 225–228.

Chakrabarti, B.K., Chakraborti, A., Chatterjee, A., 2006. Understanding and managing the future evolution of a competitive multi-agent population. Econophysics and Sociophysics: Trends and Perspectives. Wiley-VCH Verlag GmbH & Co. KGaA, Weinheim.

Chakrabarti, B.K., 2006. A Brief History of Economics: An Outsider's Account. (New Economic Windows). Springer, Milan, pp. 219–224.

Chakraborti, A.S., Chakraborti, B.K., 2009. Microeconomics of the ideal gas like market models. Phys. A 388, 4151–4158.

Challet, D., Marsili, M., Zecchina, R., 2000. Comment on thermal model for adaptive competition in a market. Phys. Rev. Lett. 85 (23), 5008.

Challet, D., Marsili, M., Zhang, Y.C., 2001. Stylized facts of financial markets and market crashes in minority games. Phys. A 294 (3–4), 514–524.

Coelho, R., Néda, Z., Ramasco, J.J., Santos, A.M., 2005. A family-network model for wealth distribution in societies. Phys. A: Stat. Mech. Appl. 353 (1–4), 515–528.

Coronel-Brizio, H.F., Hernández-Montoya, A.R., 2005. Asymptotic behavior of the daily increment distribution of the IPC, the Mexican stock market index. Revista Mexicană de Fisica 51 (1), 27–31.

Corso, G., Lucena, L.S., Thomé, Z.D., 2003. Are social structures determined by the economy? Int. J. Mod. Phys. C 14 (1), 73–80.

Defilla, S., 2007. A natural value unit-econophysics as arbiter between finance and economics. Phys. A: Stat. Mech. Appl. 382 (1), 42–51.

De Groot, E.A., Franses, P.H., 2005. Cycles in basic innovations. Econometric Institute Report 2005-35, Erasmus University Rotterdam. <http://repub.eur.nl/res/pub/6942/InnovationsCycles>. (accessed 01.10.2012).

De Groot, B., Franses, P.H., 2006. Stability through cycles. Econometric Institute Report 2006-7, Erasmus University Rotterdam. <http://repub.eur.nl/res/pub/7666/ei%202006-07> (accessed 01.10.2012).

Dorogovtsev, S.N., Mendes, J.F.F., 2002. Evolution of networks. Adv. Phys. 51 (4), 1079–1187.

Drăgulescu, A., Yakovenko, V.M., 2000. The statistical mechanics of money. Eur. Phys. J. B 17, 723–729.

Drăgulescu, A., Yakovenko, V.M., 2001. Exponential and power-law probability distributions of wealth and income in the United Kingdom and the United States. Phys. A 299, 213–221.

Ehrenstein, G., Westerhoff, F., Stauffer, D., 2005. Tobin tax and market depth. Quant. Fin. 5 (2), 213–218.

Fan, Y., Li, M., Chen, J., Gao, L., Di, Z., Wu, J., 2004. Network of econophysicists: a weighted network to investigate the development of econophysics. Int. J. Mod. Phys. B 18 (17–19), 2505–2511.

Foley, D.K., 1994. A statistical equilibrium theory of markets. J. Econ. Theory 62, 321–345.

Foley, D.K., 1996. Statistical equilibrium in a simple labor market. Metroeconomica 47, 125–147.

Fujiwara, Y., Souma, W., Aoyama, H., Kaizoji, T., Aoki, M., 2003. Growth and fluctuations of personal income. Phys. A: Stat. Mech. Appl. 321 (3–4), 598–604.

Gabaix, X., Gopikrishnan, P., Plerou, V., Stanley, E.H., 2008. Quantifying and understanding the economics of large financial movements. J. Econ. Dyn. Control 32 (1), 303–319.

Galam, S., 2003. Global physics: from percolation to terrorism: guerrilla warfare and clandestine activities. Phys. A 330, 139–149.

Galam, S., 2005. Heterogeneous beliefs, segregation, and extremism in the making of public opinions. Phys. Rev. E Stat. Nonlin. Soft Matter Phys. 71 (4), 1–5.

Gatti, D.D., Di Guilmi, C., Gaffeo, E., Gallegati, M., 2004. Bankruptcy as an exit mechanism for systems with a variable number of components. Phys. A: Stat. Mech. Appl. 344 (1–2), 8–13.

Georgescu-Roegen, N., 1971. The Entropy Law and the Economic Process. Harvard University Press, Cambridge, MA.

Gligor, M., Ausloos, M., 2007. Cluster structure of EU-15 countries derived from the correlation matrix analysis of macro-economic index fluctuations. Eur. Phys. J. B 57, 139–146.

Gligor, M., Ausloos, M., 2008a. Convergence and cluster structures in EU area according to fluctuations in macroeconomic indices. J. Econ. Integrat. 23, 297–330.

Gligor, M., Ausloos, M., 2008b. Clusters in weighted macroeconomic networks: the EU case. Introducing the overlapping index of GDP/capita fluctuation correlations. Eur. Phys. J. B 63, 533–539.

Guevara, E.H., 2006. The Why of the applicability of Statistical Physics to Economics. physics/0609088. <http://cdsweb.cern.ch/record/983169/files/0609088.pdf> (accessed 01.10.2012).

Guevara, E.H., 2007. Quantum econophysics. In: Proceedings of the Quantum Interaction AAAI Spring Symposia Series, Eds. Stanford University, Palo Alto, American Association of Artificial Intelligence, Technical Report SS-07-08, pp. 158–165. <http://arxiv.org/pdf/physics/0609245.pdf> (accessed 01.10.2012).

Guevara, E.H., 2007. Quantum games entropy. Phys. A: Stat. Mech. Appl. 383 (2), 797–804.

Haldane, A.G., May, R.M., 2011. Systemic risk in banking ecosystems. Nature 469, 351–355.

Hegselmann, R., 2004. Opinion dynamics insights by radically simplifying models. Laws Model. Sci., 1–29.

Hegselmann, R., Krause, U., 2002. Opinion dynamics and bounded confidence models, analysis, and simulation. J. Artif. Soc. Social Simul. 5 (3).

Helbing, D., Seidel, T., Lämmer, S., Peters, K., 2006. Self-organization principles in supply networks and production systems. In: Chakrabarti, B.K., Chakraborti, A., Chatterjee, A. (Eds.), Econophysics and Sociophysics: Trends and Perspectives. Wiley-VCH, Berlin.

Helbing, D., Johansson, A., Al-Abideen, H.Z., 2007. Dynamics of crowd disasters: an empirical study. Phys. Rev. E 75 (4).

Hohnisch, M., Pittnauer, S., Solomon, S, Stauffer, D., 2005. Socioeconomic interaction and swings in business confidence indicators. Phys. A 345, 646–656.

Inoue, J., 2006. Emergence of memory in networks of non-linear units: from neurons to plant cells. In: Chakrabarti, B.K., Chakraborti, A., Chatterjee, A. (Eds.), Econophysics and Sociophysics: Trends and Perspectives. Wiley-VCH, Berlin.

Ising, E., 1925. Beitrag zur theorie des ferromagnetismus. Zeitschrift für Physik 31, 253–258.

Jensen, M.H., Johansen, A., Petroni, F., Simonsen, I., 2004. Inverse statistics in the foreign exchange market original research article. Phys. A: Stat. Mech. Appl. 340 (4), 678–684.

Kar Gupta, A., 2006. Models of wealth distributions—a perspective. In: Chakrabarti, B.K., Chakraborti, A., Chatterjee, A. (Eds.), Econophysics and Sociophysics: Trends and Perspectives. Wiley-VCH Verlag GmbH & Co. KGaA, Weinheim.

Kenett, D.Y., Raddant, M., Zatlavi, L., Lux, T., Ben-Jacob, E., 2012. Correlations and dependencies in the global financial village. Abstract posted by Dror Kenett on <http://www.unifr.ch/econophysics/>, September 2012.

Lamper, D., Howison, S., Johnson, N.F., 2008. Predictability of large future changes in a competitive evolving population. <http://arxiv.org/pdf/cond-mat/0105258.pdf> (accessed 01.10.2012).

Latora, V, Marchiori, M., 2003. The Architecture of Complex Systems. Santa Fe Institute for Studies of Complexity, Oxford University Press, Oxford.

Latora, V., Marchiori, M., 2004. How the science of complex networks can help developing strategies against terrorism. Chaos Solitons Fractals 20, 69−75.

Liossatos, P.S., 2004. Statistical Entropy in General Equilibrium Theory. Working Papers 0414, Florida International University. Available at <http://casgroup.fiu.edu/pages/docs/2245/1280267976_04-14.pdf> (accessed 01.10.2012).

Mantegna, R.N., Stanley, H.E., 1997. Econophysics: scaling and its breakdown in finance. J. Stat. Phys. 89 (1−2), 469−479.

Mantegna, R.N., Stanley, H.E., 2000. An Introduction to Econophysics: Correlations and Complexity in Finance. Cambridge University Press, Cambridge.

Mantegna, R.N., Palágyi, Z., Stanley, H.E., 1999. Applications of statistical mechanics to finance. Phys. A: Stat. Mech. Appl. 274 (1), 216−221.

Marsili, M., Maslov, S., Zhang, Y.C., 1998. Comment on "Role of intermittency in urban development: A model of large-scale city formation". Phys. Rev. Lett. 80 (21), 4830.

Marsili, M., Challet, D., 2001. Trading Behavior and Excess Volatility in Toy Markets. Advances in Complex Systems, 4 (01), 3−17 (World Scientific).

Maslov, V.P., 2002a. Econophysics and quantum statistics. Math. Notes 72 (5−6), 811−818.

Maslov, V.P., 2002b. The notions of entropy, hamiltonian, temperature, and thermodynamical limit in probability theory used for solving model problems in econophysics. Russ. J. Math. Phys. 9 (4), 437−445.

Matsushita, R., Gleria, I., Figueiredo, A., Da Silva, S., 2007. Are pound and euro the same currency? Phys. Lett. Section A Gen. At. Solid State Phys. 368 (3−4), 173−180.

McCauley, J.L., 2004a. Dynamics of Markets: Econophysics and Finance. Cambridge University Press, Cambridge.

McCauley, J.L., 2004b. What economists can learn from physics and finance. <http://mpra.ub.uni-muenchen.de/2240/1/MPRA_paper_2240.pdf> (accessed 01.10.2012).

Mehta, A, Majumdar, A.S., Luck, J.M., 2005. How the rich get richer. In: Chatterjee, A., et al., (Eds.), Econophysics of Wealth Distributions. Springer, Italia, pp. 199−204.

Mimkes, J., 2000. Society as a many-particle system. J. Therm. Anal. 60 (3), 1055−1069.

Miśkiewicz, J., Ausloos, M., 2005. Correlations between the most developed (G7) countries. A moving average window size optimization. Acta Phys. Pol. B 36, 2477−2486.

Miśkiewicz, J., Ausloos, M., 2006a. An attempt to observe economy globalization: the cross correlation distance evolution of the top 19 GDPs. Int. J. Mod. Phys. C 17 (3), 317−331.

Miśkiewicz, J., Ausloos, M., 2006b. G7 country gross domestic product (GDP) time correlations—a graph network analysis. In: Takayasu, H. (Ed.), Practical Fruits of Econophysics. Springer, Tokyo, pp. 312−316.

Mizuno, T., Katori, M., Takayasu, H., Takayasu, M., 2001. Statistical Laws in the Income of Japanese Companies. <http://arxiv.org/pdf/cond-mat/0308365> (accessed 01.10.2012).

Mizuno, T., Takayasu, M., Takayasu, H., 2002. The mechanism of double-exponential growth in hyper-inflation. Phys. A: Stat. Mech. Appl. 308 (1−4), 411−419.

Mizuno, T., Takayasu, H., Takayasu, M., 2006. Correlation networks among currencies. Phys. A: Stat. Mech. Appl. 364, 336−342.

Mostardinha, P., Durana, E.J., Vistulo De Abreu, F., 2006. The econophysics in the euromillions lottery. Eur. J. Phys. 27 (3), 675−684.

Newman, M.E.J., 2004. Who is the best connected scientist? A study of scientific coauthorship networks. In: Ben-Naim, E., Frauenfelder, H., Toroczkai, Z. (Eds.), Complex Networks. Springer, Berlin, pp. 337–370.

Ohtaki, Y., Hasegawa, H.H., 2003. Superstatistics in Econophysics. <http://arxiv.org/pdf/cond-mat/0312568.pdf> (accessed: 02.10.12.)

Olivera, N.L., Proto, A.N., Ausloos, M., 2011. Information society: modeling a complex system with scarce data. J. Proc. V Meet. Dyn. Soc. Econ. Syst. 6, 443–460.

Palmer, J., 2007. Planar Ising Correlations. Birkhäuser, Boston, MA.

Reimann, S., 2006. The process of price formation and the skewness of asset returns. Working Paper Series/Institute for Empirical Research in Economics, No. 276, University of Zurich. <http://www.zora.uzh.ch/52233/1/iewwp276.pdf> (accessed 01.10.2012).

Richmond, P., Hutzler, S., Coelho, R., Repetowicz, P., 2006. A review of empirical studies and models of income. In: Chakrabarti, B.K., Chakraborti, A., Chatterjee, A. (Eds.), Econophysics and Sociophysics: Trends and Perspectives. Wiley-VCH, Berlin.

Roehner, B.M., 2005. Patterns of Speculation: A Study in Observational Econophysics. Cambridge University Press, Cambridge.

Săvoiu, G., Iorga Simăn, I., Crăciuneanu, V., 2012. The phenomenon of concentration—diversification in contemporary globalization. Rom. Stat. Rev. 4, 16–28.

Scafetta, N., West, B.J., Picozzi, S., 2004. A trade-investment model for distribution of wealth. Phys. D: Anomalous Distrib. Nonlin. Dyn. Nonextensivity 193, 338–352.

Scalas, E., 2005. Basel II for Physicists: A Discussion Paper. <http://arxiv.org/abs/cond-mat/0501320v1.pdf> (accessed 01.10.2012).

Scalas, E., 2006. The application of continuous-time random walks in finance and economics. Phys. A: Stat. Mech. Appl. 362 (2), 225–239.

Scalas, E., Gallegati, M., Guerci, E., Mas, D., Tedeschi, A., 2006. Growth and allocation of resources in economics: the agent-based approach. Physica A 370, 86–90.

Schulze, C., 2002. The domino effect for markets. Int. J. Mod. Phys. C 13 (2), 207–208.

Sen, P., 2006. Complexities of social networks: A physicist's perspective. In: Chakrabarti, B.K., Chakraborti, A., Chatterjee, A. (Eds.), Econophysics and Sociophysics: Trends and Perspectives. Wiley-VCH, Berlin.

Sinha, S., Kumar Pan, R., 2006. How a 'hit' is born: The emergence of popularity from the dynamics of collective choice. In: Chakrabarti, B.K., Chakraborti, A., Chatterjee, A. (Eds.), Econophysics and Sociophysics: Trends and Perspectives. Wiley-VCH, Berlin.

Situngkir, H., 2004. The political robustness in Indonesia. Evaluation on hierarchical taxonomy of legislative election results 1999 and 2004. <http://cogprints.org/3625/1/robustness_of_politics_ind.pdf>.

Situngkir, H., 2006. Advertising in duopoly market. Working Paper BFI No. WPG2006. <http://mpra.ub.uni-muenchen.de/885/1/MPRA_paper_885.pdf>.

Situngkir, H., Surya, Y., 2004. Agent-based model construction in financial economic system. Working Paper Bandung Fe Institute. <http://cogprints.org/3767/1/hokky_new2004.pdf> (accessed 01.10.2012).

Smith, D.M.D., Johnson, N.F., 2006. Understanding and managing the future evolution of a competitive multi-agent population. In: Chakrabarti, B.K., Chakraborti, A., Chatterjee, A. (Eds.), Econophysics and Sociophysics: Trends and Perspectives. Wiley-VCH, Berlin.

Smith, E., Foley, D.K., 2008. Classical thermodynamics and economic general equilibrium theory. J. Econ. Dyn. Control 32, 7–65.

Solomon, S., Richmond, P., 2001. Stability of Pareto—Zipf Law in Nonstationary Economies. In: Kirman, A., Zimmermann, J.-B. (Eds.), Economics with Heterogeneous Interacting Agents, Springer, Heidelberg. (Lecture notes in Economics and Mathematical Systems), 503, pp. 141–159.

Sousa, T., Domingos, T., 2006. Equilibrium econophysics: a unified formalism for neoclassical economics and equilibrium thermodynamics. Phys. A: Stat. Mech. Appl. 371 (2), 492–512.

Stauffer, D., Schulze, C., 2005. Microscopic and macroscopic simulation of competition between languages. Phys. Life Rev. 2 (2), 89–116.

Stinchcombe, R., 2006. Zero-intelligence models of limit-order markets. In: Chakrabarti, B.K., Chakraborti, A., Chatterjee, A. (Eds.), Econophysics and Sociophysics: Trends and Perspectives. Wiley-VCH Verlag GmbH & Co. KGaA, Weinheim.

Stanley, E., 2008. Econophysics and the current economic turmoil. The Back Page, APS (American Physical Society) News, 17 (11), 8–9.

Stanley, E.H., Gabaix, X., Gopikrishnan, P., Plerou, V., 2006. Economic fluctuations and statistical physics: the puzzle of large fluctuations. Nonlin. Dyn. 44, 329–340.

Takahashi, T., 2006. A mathematical framework for probabilistic choice based on information theory and psychophysics. Med. Hypotheses 67 (1), 183–186.

Tsallis, C., 2009. Introduction to Nonextensive Statistical Mechanics—Approaching a Complex World. Springer, New York, NY.

Tsallis, C., Anteneodo, C., Borland, L., Osorio, R., 2003. Nonextensive statistical mechanics and economics. Phys. A 324, 89, http://arxiv.org/pdf/cond-mat/0301307. (accessed 01.10.2012).

Wang, Y., Xi, N., Ding, N., 2006. The contribution of money transfer models to economics. In: Chakrabarti, B.K., Chakraborti, A., Chatterjee, A. (Eds.), Econophysics and Sociophysics: Trends and Perspectives. Wiley-VCH, Berlin.

Watanabe, A., Uchida, N., Kikuchi, N., 2006. Econophysics of precious stones. <http://arxiv. org/pdf/physics/0611130v1.pdf> (accessed 01.10.2012).

Weisbuch, G., Battiston, S., 2007. From production networks to geographical economics. J. Econ. Behav. Organ. 64 (3–4), 448–469.

Weisbuch, G., Deffuant, G., Amblard, F., 2005. Persuasion dynamics. Phys. A 353, 555–575.

Westerhoff, F.H., 2005. Consumer behavior and fluctuations in economic activity. Adv. Complex Syst. 8 (2–3), 209–215.

Willis, G., 2005. Laser welfare: first steps in econodynamic engineering. Phys. A 353, 529–538.

Yakovenko, V.M., 2009. Econophysics, statistical mechanics approach to. In: Meyers, R.A. (Ed.), Encyclopedia of Complexity and System Science. Springer, Berlin.

Yakovenko, V.M., 2011. Statistical mechanics approach to the probability distribution of money. In: New Approaches to Monetary Theory: Interdisciplinary Perspectives. Heiner Ganssmann, Oxford, Routledge.

Yakovenko, V.M., Rosser, J.B., 2009. Colloquium: statistical mechanics of money, wealth, and income. Rev. Mod. Phys. 81, 1703–1725.

Zarama, R., et al., 2007. Rethinking research management: Colombia. Kybernetes 36, 364–382.

Zimmermann, R.E., Soci, A., 2003. The emergence of Bologna and its future consequences. Decentralization as cohesion catalyst in guild dominated urban networks. <http://arxiv.org/pdf/cond-mat/0411509v1.pdf> (accessed 01.10.2012).

Zimmermann, R.E., Soci, A., Colacchio, G., 2001. Re-constructing Bologna. The city as an emergent computational system. An interdisciplinary study in the complexity of urban structures. <http://arxiv.org/pdf/nlin.AO/0109025> (accessed 01.10.2012).

PART *III*

Sociophysics

CHAPTER *8*

A Physical Model to Connect Some Major Parameters to be Considered in the Bologna Reform

Radu Chişleag
Bucharest Polytechnic University, Faculty of Applied Sciences, Romania

8.1 INTRODUCTION

The Bologna Reform Process includes major objectives such as enlarging democratic access to higher education, better structuring it, increasing academic excellence to become competitive in the knowledge society and in the world competition, and eventually, not increasing the public funding too much.

Here a simple physical model is developed, which permits to emphasize the relationships between these quantities (objectives), of different dimensions, in connection with other quantities (parameters), which are important in higher education. Each quantity is defined so as to easily understand its physics dimensions, and to be easily quantified, for the later development and use of the physical model by sociologists and economists.

For the purpose of this chapter, in order to avoid getting an overcharged model, we shall suppose a temporary quasistability (small relative values of the derivatives) in the evolution: of the number of population, of its age and education distribution, and of its academic behavior, during education. In this chapter, the model will only be exploited qualitatively. Later on, this model may be developed so as to include yearly and specific variations, and fed with quantified values of the considered quantities, acquired from field studies, or literature research (Chişleag, 2008).

8.2 THE PHYSICAL MODEL

Let us consider the society we refer to (region, country, state, federation, union) as having a population of P members and a class population (cohort) of Pe, representing that part of population born along 1 year, at an age corresponding to the normal age of enrolment in higher education (HE):

$$\text{Let pe, equal to pe} = Pe/P \tag{8.1}$$

be referred to as the proportion of population at the normal age of enrolment in HE. Let Sne be the number of newly enrolled students in the HE institutions of the considered society, during 1 academic year. Sne, the number of newly enrolled students, may include initial students and adult students. The quantity a equal to:

$$a = \text{Sne}/\text{Pe} \tag{8.2}$$

could be defined as the rate of access to enrolment in HE (shortly—academic access).

Therefore, Sne is equal to

$$\text{Sne} = a * \text{Pe} = a * \text{pe} * N \tag{8.3}$$

Let d $(0 < d < 1)$ be the dropout rate for 1 year of study. Let f $(0 < f < 1)$ be the failure rate during 1 year of study. Let Sbe be defined as the total number of students at the beginning of the first academic year.

Because of the dropout, the total number of students remaining enrolled at the end of that academic year, Seed, is:

$$\text{Seed} = (1 - d) * \text{Sbe} \tag{8.4}$$

Because of the rate of failure being present, too, the total number of students at the beginning of the first year becomes

$$\text{Sbe} = [1 + f(1 - d)] * \text{Sne} = [1 + f(1 - d)] * a * \text{Pe} = [1 + f(1 - d)] * a * \text{pe} * N \tag{8.5}$$

Now, considering the effects of the dropout and failure, in the first approximation, the number of equivalent students at the end of 1 year of study, Seedf, is equal to

$$\text{Seedf} = (1 - d) * [1 + f(1 - d)] * \text{Sne} = (1 - d) * [1 + f(1 - d)] * a * \text{pe} * N$$
$$(8.6)$$

smaller than the number of totally financed students in that year of study.

The number of students graduating during 1 academic year, Sg, is smaller than Sne (at the beginning of the cycle), due to the dropout (here considered of constant rate, for each of the T years of study):

$$\text{Sg} = \text{Sne} * (1 - d)T < \text{Sne} \qquad (8.7a)$$

If the dropout rate, d, is small (like for prestige universities), we might approximate Eq. (8.7a) with

$$\text{Sg} = \sim \text{Sne} * (1 - Td) \qquad (8.7b)$$

The rate of access to graduation, ag, may be defined as

$$\text{ag} = \text{Sg}/\text{Pe} = \text{Sg}/\text{peN} = a * (1 - d)T < a \qquad (8.8)$$

For a T years cycle, because of failures, the average actual duration of study of a graduate is not more than T, but Ta is approximately equal to

$$\text{Ta} = T * (1 + f)R > T \qquad (8.9a)$$

where R is the number of yearly failures permitted by the existing academic regulations.

For a small failure rate, f, we may write a linear relationship for (8.9a):

$$\text{Ta} = \sim T * (1 + R * f) \qquad (8.9b)$$

For a cycle of T years normal length, the number of financed years of study per graduate, Tfg, is not greater than T but, considering both the rates of failure f and of dropout d (i.e. the financing of students who are failing and/or not graduating) covered by the society (public funding), it is

$$\text{Tfg} = T * (1 + f)R * (1 - d) - L \qquad (8.10a)$$

where L is the length in years and the dropouts be financed by the education system.

For small f and d:

$$\text{Tfg} = \sim T * (1 + R * f) * [1 + L * d] > \text{Ta} > T \qquad (8.10b)$$

The society does not financially cover the failure (like in Romania, in recent years) but cover the dropout for 1 year ($L = 1$):

$$\text{Tfg} = T * (1 - d) - L = T * (1 - d) - 1 \qquad (8.11a)$$

respectively, for a small d,

$$\text{Tfg} = \sim T * [1 + d] > \text{Ta} > T \qquad (8.11b)$$

We have to accept that, in democratic societies, where there is a requirement that all students have to comply with the same academic standards and the admission selection is based mainly upon merit (and even upon excellence for a few universities), an increase in the academic access rate, a, would probably mean an increase in both f, the rate of failure, and d, the dropout rate.

More, because the increase in the academic access a is meant to address not only initial but also adult students; the new segment of the adult students is to probably do more nonacademic work during studies than the initial students, so as to ensure the living expenses for them and sometimes for their dependents, and be academically less efficient (due to lack of continuity, too); this, again, meaning an increase in both failure, f, and dropout d rates, when increasing a.

Consequently, Tfg is rather strongly increasing with the increase of a and correspondingly, ag/a is, in turn, decreasing when a is increasing.

The total number of students enrolled and financed in a T-year cycle, Sc is

$$\text{Sc} = \text{Sg} * \text{Tfg} > \text{Sg} * \text{Ta} > \text{Sg} * T \qquad (8.12)$$

Thus, enlarging democratic access to higher education (essentially accompanied by the increasing in failure and dropout rates) means increasing of Sc and Tfg and, consequently, a larger relative increase in the number of teachers than in the number of the rate of access.

Thus HE may be considered as a generator of demand of services (of education) and consequently a generator of jobs, rather than a consumer of public resources. Therefore, the perception of the social effect of increasing the rate of access might, eventually, be positive. Let c be referred as the unit cost and defined as the average cost of 1 year of study for one student enrolled in that cycle.

It has to be mentioned here that this cost of educating one student during 1 academic year, $c = c(a)$, is probably to increase with the increase of a, because of the new investments in equipment, buildings, and human resources (educators) necessary to offer more students higher education than the actual academic infrastructure allows.

But, in those societies where the actual academic capacities are not entirely used, the fixed expenses might not increase and $c(a)$ might slightly decrease with a, at least until the maximum existing capacities are reached.

The society's budget necessary to run a cycle, B, is

$$B = Sc * c = Sg * Tfg * c = Sg * c * T * (1 + f)R * (1 - d) - L$$
$$= \sim Sgc * T * (1 + R * f) * [1 + L * d] \tag{8.13}$$
$$= \sim ag * T * c * (1 + R * f) * [1 + L * d] * pe * N$$

The cost per graduated student would be

$$B/Sg = \sim c * T * (1 + R * f) * [1 + L * d] = c * Tfg \tag{8.14}$$

much larger than $c*T$ and fast increasing with f, d, c (and with a and ag).

Thus, the decision to increase access to a cycle of higher education, without other measures, will require an increase in the budget for running a cycle, the relative increase of B being much important than the relative increase of the access to that cycle of HE, this correlation generating a large effort for the public budget.

This conclusion is to be completely understood by the politicians deciding a larger democratic access to higher education and by the members of the society who are paying taxes but benefiting, too, from such a decision, not to consider here the workers abroad, not contributing to the education budget at home. But, usually, parliaments and/ or governments do not easily accept the increase of quota of higher education in the public budget, in spite of the possibility of thus generating new jobs in education with a low specific investment (the investment for creating and maintaining a job in HE being much smaller than in the major part of new industries).

The deduced relationship (8.13) clearly explains one feature of the essence of the Bologna Declaration: no increase in the academic access is possible in the current budgetary frame, without a structural reform of higher education.

To run all three cycles: I (B.Sc.); II (M.Sc.), and III (Ph.D.) ($O =$ I, II, III), the necessary budget for higher education, BHE, would be

$$\begin{aligned}
\text{BHE} = \text{BI} + \text{BII} + \text{BIII} &= \text{pe} * N * [(\text{Tfg} * \text{ag} * c)\text{I} + (\text{Tfg} * \text{ag} * c)\text{II} \\
&\quad + (\text{Tfg} * \text{ag} * c)\text{III}] \\
&= \text{pe} * N * SO[T * (1 + f)R * (1 - d)T - 1 * \text{ag} * c]O \\
\sim &= \text{pe} * N * ScO\{T * (1 + R * f) * [1 + (T - 1)d] * \text{ag} * c\}O
\end{aligned}$$

$$(8.15)$$

The component parentheses differ significantly between them by:

extensive parameters:

- (e1) rate of access to higher education, a;
- (e2) duration of study T;
- (e3) eventually, by the number of cycles O, financed for one individual and by

intensive parameters:

- (i1) failure rate, f;
- (i2) dropout rate, d (f and d influencing the rate of access to graduation ag);
- (i3) unit cost per years of study, c.

An increase in every one of these parameters results in an increase of the budget for higher education.

Let F be the total yearly funding of higher education, where FP comes from public sources, FC comes from private sources (companies), FU comes from the direct revenues of the universities due to services done by them (including taxes). Then:

$$F = FP + FC + FU \qquad (8.16)$$

Among the majority of EU member states, F, the funding of higher education is ensured mainly by public contribution, with the exception of a part of doctoral costs and a small part of master costs, which are covered by private companies. In all Central and East European (CEE) former communist countries, the public funding, only, practically, counts.

A balanced budget of the HE means:

$$BHE = F \qquad (8.17)$$

For the needs of this model, academic excellence, E, might be defined as the quality and the quantity of the new personal knowledge and abilities acquired by a graduate of a cycle, by the new knowledge and new types of products generated by one student of a cycle during his or her graduation studies.

Sociological and psychological investigations have shown that E is very fast increasing (some would say quasiexponentially) with:

- quality and the volume of the basic knowledge K;
- intelligence quotient Q (or something similar to IQ, but corresponding to the student age);
- creativity, G,
- motivation, M,
- desire to win, V
- habit of working, L.

Training for research age, A (for small A), A being the time span since starting academic education and being generated the first genuine results.

- Supplementary investment, I, for training for excellence (investment in equipment, supervising, organizing competitions, etc). When expressed as a function of the previous parameters, excellence is

$$E = EMREF * \exp(bKK + bQQ + bGG + bMM + bLL + bAA + bII) \qquad (8.18)$$

where EMREF is a reference value of academic excellence, corresponding to some defined levels of values for the considered parameters (e.g., to the minimum requirements for enrolment) and bs are system constants to be determined.

This function E is a very sharply increasing one, with the increase of each of the parameters determining the excellence. It may be found in sociological and psychological reports that the intensity (value) of all parameters K,..., I are maximum decreasing when going further from the top values among the members of a society, that is with increasing the rate of access to higher education, a, especially in democracies, where the education is free and the selection is based on merit. In a first approximation that leads to a relationship of the form

$$E = \text{EMREF} * \exp(-a) \tag{8.19}$$

where EMREF is a high reference value of academic excellence, to be found considering, for example, the top $10\text{-}4*N$ segment of the population of the considered society.

Because of the negative derivative of E with respect to a, the increase in the democratic access, a, will induce the decrease in the average individual academic excellence, in spite of the necessary increasing funding.

Nevertheless, the society's cumulated academic excellence, ET, the total quantity of new knowledge, and new types of products generated by higher education will be slightly improved, but at a higher unit cost, the newly generated cumulated academic excellence obtained by enlarging academic access being more expensive, larger the increase in access be. What has to be done to increase the access a to higher education, subject to social restrictions (with reference based upon the mentioned relationships implied in the developed model)? Relations (8.13) and (8.16) give some hints to how to increase a (an intensive parameter) without increasing the quota of the budget of HE in the society's budget: to reduce the values of the other extensive parameters and to reduce the values of the intensive parameters. But, until very recently, the actual structural reform proposed by politicians was a rather extensive one. The major recommendations done and actions taken are mainly concerned with acting on the extensive parameters, the intensive factors being applied to, rather partially and rarely, only as auxiliary ones when changing the extensive factors.

What has been recommended is:

1. To reduce the number O of cycles of HE, by criticizing the variety of types and large number of programs in higher education (until recently).
2. To have a three-cycle structure, to be generalized all over Europe.
3. To reduce the number of years T spent by an individual during his higher education.
4. Particularly in each cycle.
5. Especially in the first cycle.
6. By recommending the division of the traditional integrated first cycle (particularly in engineering education) in two cycles: bachelor and master.

To increase a does mean for some politicians, in fact, to increase aI, the academic access to the first cycle, mainly, and consequently to reduce the effective duration of the first cycle TI—see Eq. (8.13).

That explains the Bologna Declaration recommendation to generalize a short first cycle, by renouncing the integrated long cycle. One or even 2 years out of five, less in an integrated first 5-year cycle, means a possible increase of 20–40% in the rate of academic access aI on that segment of curricula! The action upon the extensive parameters may be improved, too, by considering the specific reduction of the intensive parameter c, the unit yearly cost:

Because the unit costs c increases with the order of the cycle,

$$cI < cII < cIII \tag{8.20}$$

one may increase a by preserving the actual BHE, not only by reducing the duration T of each cycle but, eventually, by changing the distribution of students among cycles, toward lower unit costs, that means to reduce aII, the rate of access for the second cycle, mainly by separating it from a first (integrated) cycle and the saved money to be transferred to the first cycle to increase aI, more than the reduction in the aII, so increasing the sum aI + aII. Unit cost c may be reduced by intensive use of information technologies.

Continuous education may reduce the public funding, FP, by increasing the direct revenues of universities, FU, through charging the adult students with the fees or the private companies for (eventually specialized or vocational) education. All these measures have been recommended in the Bologna Declaration process. There has been some reluctance to implement these measures, not only in the academic media, which accuse these recommendations of harming academic excellence, because of compressing 4 or 5 year teaching in a 3-year program, but also in the industrial media, which accuse these recommendations of not offering enough specialized graduates to comply with the (private) industry needs.

Both directions of criticism might be understood, but they are rather tributary to the past (academic inertia on the content of education, neglecting the role of improving intensive factors, f and d, for example) and to the (unjustified) fear of there being lost jobs in education, or tributary to private peculiar interests—the hidden requirement related to public funding needed to cover entirely the specialized education and even to educate (train) the specialists needed by private companies in exactly their fields of interest and exactly in the quantities required presently, without their contribution. The recent multilateral and bilateral decisions to launch large projects to stimulate research in Europe may be understood as a stimulating factor to increase excellence, therefore to act upon the intensive factors as defined in this model.

What might be done? To act upon intensive parameters and to better manage the human assets! Some actions can be suggested by using the model herein developed:

1. An improvement of the quality of the academic work and more of the work during previous stages of education (school). A stimulation of the motivation would result in smaller f and d and might have an important role in enlarging successful access to higher education with not very large public efforts for financially sustaining it, even when not changing the extensive parameters of the HE. This improvement of the quality of the academic work and more of the work during previous stages of education (school) will have a positive influence on academic excellence, too, and may contribute to not overcharging the society's budget.

2. A solution which seems convenient is to have two different types of the first short cycles:

   a. one as a rather unique cycle, offering a mixture of general and specialized HE to students intending to become routine workers and officers;

   b. the second type, a first cycle as a part of at least two cycles, the first cycle being devoted to general basic education with the specialized education being ensured by the second cycle.

The stress on general education in this first cycle will have effects, not only in the middle and long runs, upon the process of education, but also in the short run, because general education is cheaper than specialized education; may be done in larger classes; may benefit more from IT and would benefit from scale savings (this being difficult in the case of specialized education). Besides, a good general education will be a healthy basis in ensuring an improvement of academic excellence, and this goal being reached by reducing the unit average costs.

It is known that only a small part of the graduates who are hired after graduation find jobs in their specialization field. Thus, public money is spent without enough reason, for specialization in the first cycle, or in an integrated cycle (like, before Bologna, for polytechnic universities), in spite of the fact that general education is required for successful hiring.

3. To fund general education from public sources and specialized education, more than now, by private funding.

4. To organize targeted specialization programs with curricula on topics of present or of forecasted interest for companies, together with the companies concerned. These companies are to ensure private funding, to complement public funding. By such increasing of FC, the public funding FP II may be, eventually, not increased, maintained at the actual level or even decreased, thus reducing the burden for the society's budget.

5. Therefore, hiring specialists graduating from universities might be considered by private companies as a type of "outsourcing" permitting them to fund universities for the education of exactly the specialists the companies require.

6. This specialization might be done under contracts signed by the interested applying student, binding themselves to work for the sponsor, a determined period of time, or else to pay back the academic fees, the way the armed forces do in some countries.

7. To increase ET (cumulated excellence), a different approach might be necessary when increasing a, from the linear one, the only one usually considered by many decision-makers.

8. This new approach should be suited to the exponential format of E—see Eqs (8.18) and (8.19) and the hierarchy in capability of the students be detailed by a specific approach for each class of scarcity of gifted individuals.

9. Thus, for the needs of this model, we may consider the students ranked in function of the logarithmic decrement of their academic excellence (the logarithm with changed sign of the relative frequency of the segment of respective individuals in society).

10. Thus, we may have a hierarchic structure. This hierarchic structure may be described using the logarithmic base 10 and having the classes of scarcity:

   a. Normal students: the first class of logarithmic decrement $H = 1$; relative frequency: $10^{-1}-10^{-0.6}$ ("upper half").

   b. Gifted students: $H = 2$; relative frequency: $10^{-2}-10^{-1}$.

   c. Very gifted students: $H = 3$; relative frequency: $10^{-3}-10^{-2}$.

   d. Excellent (at national level) students: $H = 4$; relative frequency: $10^{-4}-10^{-3}$.

   e. World value students: $H = 5$; relative frequency: $10^{-5}-10^{-4}$.

   f. Historical value students: $H = 6$; relative frequency: $10^{-6}-10^{-5}$.

11. The individuals belonging to each class of value, H, are to be selected as early as possible, for each class H, from the smaller H classes, starting with the selection of class $H = 2$, from the basement (normal class), and so on, increasing H gradually.

12. This selection is correct and permanently needs a system of promotions into a higher class, adequately designed and implemented.

13. But, what is more important is to work adequately with the selected segment of students so as to be able to produce as much excellence as possible, upon which to select the upper class of excellence H as soon as possible.

   The entire bee hive work for the queen of the bees!

14. These selected students are to be implied in new programs, in new fields in sharp progress. The newer the programs are, the stiffer the selection is, but the higher the yield.

15. The increase of the sizes of upper H classes in a university would be an important factor in creating the own resources of the university, ensuring the success of the entrepreneurial university.

16. It may be found that, starting with the $H = 3$ class, the investment in the supplementary education for excellence be so large that only top universities and strong companies could possibly approach it, the way strong football clubs approach the top football players and trainers.

17. The next step would be to introduce "contracts for education for excellence," which are to pay selected students and bind the

so-educated people to the "educator," very much as in the top football for example.

18. A system of awards and of competitions on excellence (creation), based upon healthy criteria and using efficient procedures to increase the sense of responsibility of the students, will increase almost all of the parameters leading to excellence, mentioned above, and by this, will increase the average of academic excellence, finally resulting in the reduction of funding for the same results.

19. Probably, the system of promoting by competition will have to be assisted by a system of penalties for those not replying to the society's effort for them, like:
    (a) paying all previous academic fees when dropping out of a cycle or waiving of academic fees only for the standard duration (e.g. 3 years, for 180 credit points for example);
    (b) paying the integral academic fee for the year failure;
    (c) paying a fee for each credit point of a failed examination.
    Due to such penalties, the increase in the budget produced by f and d in Eqs (8.13) and (8.16) will be supported not by the society but by the consumers of education involved.

20. This system of awards and penalties in higher education proposed emphasizes the importance of educating better from childhood not only for getting more excellence but to get cheaper democratic access.

21. The current world top universities do not need to greatly change their behavior if increasing the democratic access, but they would benefit from selecting candidates from a larger base.

22. The existing reserves in education infrastructure must be exploited.

23. New universities, aiming to manage enlarged democratic access without a stiff reduction in academic excellence are to be designed and created.

8.3 CONCLUSIONS

A physical model to explore and explain social issues and suggest social solutions enlarges the scope of education and research (and of decision makers, too!). The proposed physical model describes well what is known as yet but offers some suggestions for ways of improving the actual situation, as well. The pre-Bologna academic system is not entirely fit to ensure both increasing academic access and preserving academic excellence. Both increasing the academic access and preserving academic excellence imply supplementary funding efforts and possibly restructuring the goals of higher education, the content, the quality of activity and of management of human assets. This restructuring has to be extended to high schools. Getting more academic excellence from each segment of the academic population would mean much more effort and possibly a different stress on the components of the academic structure: B.Sc.—to be devoted to general education; M.Sc.—to be devoted to specialized education; and Ph.D.—to be devoted to generation of new knowledge and new products. This is corresponding to the Bologna Declaration spirit. Enlarging democratic access to higher education has to be considered as a generator of demand of services, and consequently of jobs with a low specific investment and not only as a consumer of public resources. Public funding has to be oriented toward general education; private funding must be attracted for specialized education and both must contribute to doctoral education, because of their specific goals and needs.

The reduction of costs and the improvement of academic performances require larger efforts during preuniversity education, and the development of a full system of rewards and penalties and of competitions to increase motivation. The academic excellence has to be structured on three to four levels and managed differently on each level. The management of the academic structure has to be included into the criteria for academic excellence. The top academic performers (having a relative frequency in their cohort of 10-3-10-5) are to be selected at the earliest age and educated under contract, very much as the top sportsmen are trained and their careers managed. Increasing democratic academic access must be accompanied by a better and differential education of the elite, the main generator of new types of resources to basically ensure the increasing democratic academic access and the sustainable progress of mankind.

REFERENCE

Chişleag, R., 2008. A physical model meant to connect some major parameters to be considered in the Bologna reform. Res. Cent. Adv. Mater. Econophys. J. 1 (1), 7–12, http://www.ccma.upit.ro/EDEN/abstracts_EN.htm. (accessed 02.10.2012).

CHAPTER 9

Statistical Physics Models for Group Decision Making

Mircea Gligor

National College "Roman Vodă", Roman, Neamţ, Romania

9.1 INTRODUCTION

The application of theoretical physics methods to the sphere of the social sciences (a field frequently called "sociophysics") is a natural extension of econophysics. During the last decade of the twentieth century, the statistical physics tools have found many applications in the study of the economic–financial processes. The same methods proved also to be useful in the investigation of other social phenomena, including social psychology. In addition to the classical statistical physics tools, sociophysics uses frequent methods developed in nonlinear dynamics, computational sciences, evolving networks, and theoretical biophysics, pertaining in this way to the larger framework of the "complex systems sciences."

Except for several previous remarkable works (Galam, 1986; Haken, 1983; Montroll, 1987), sociophysics was extensively developed during the last decade, calling into debate various issues, phenomena, and associate models that have permanently aroused a large interest among economists, sociologists, psychologists, and anthropologists. Some fundamental topics and several pioneering works are given below:

- Hierarchical organization of the cities and nations (Malescio et al., 2000).
- Wealth condensation in the democratic societies (Bouchaud and Mézard, 2000; Drăgulescu and Yakovenko, 2000).
- Demographic issues and the dynamics of populations (Giardina et al., 2000; Gligor and Ignat, 2001).
- Group decision making and nonlinear dynamics (Gligor, 2001; Guzmán-Vargas and Hernández-Pérez, 2006; Laguna et al., 2003).
- The social impact and leader's influence in opinion formation (Holyst et al., 2000; Kacperski and Holyst, 2000; Stauffer and Sahimi, 2007).
- The political configuration of social groups (individuals, countries) and totalitarianism (Galam, 2000a, 2000b, 2002).
- The minority game and its applications in social psychology (Challet and Zhang, 1998; Mosetti et al., 2006), etc.

Some inherent difficulties of the social systems modeling are due to the fact that the functional dependences in the social sciences have to take into account circumstances that differ substantially from those encountered in the natural sciences. Firstly, experimentation is usually not feasible and is replaced by survey research, implying that the explanatory variables cannot be manipulated and fixed by the researcher. Secondly, the number of possible explanatory variables is often quite large, unlike the small number of carefully chosen treatment variables frequently found in the natural sciences, known as relevant variables. Thirdly, for many social systems, the time series are composed of data having a quarterly or at most monthly frequency; this fact implies some precautions in data processing and prediction.

In modeling the human group's behavior, a crucial point is to study the group decision making and the related issue of the collective opinion formation and dynamics. These topics are analyzed in the following two sections of the present work, starting from the pioneering models proposed by Galam (2000b) and Holyst et al. (2000). Several numerical simulations were performed in order to discuss the above mentioned models (Gligor and Ignat, 2003).

Going to larger and larger microsocial cells, and increasing, at the same time, the number of possible decisions, one gets macrosocial structures, usually called *clusters*. The cluster analysis was firstly introduced in social sciences (Tryon, 1939), and then was successfully applied in physics, chemistry, biology, and computer sciences. A large number of political, economic, social, and administrative decisions are embodied in the economic growth, currently measured by GDP and GDP per capita. Starting from the fluctuations of these indicators, one can establish correlations among the countries that adopt similar lines of development, in a given time interval. Section 9.4 analyzes the possibility of mapping the correlation coefficients between the macroeconomic time series in weights attached to the links of a fully connected graph, getting in this way a *weighted network*. This structure pertains to the special classes of growing networks of special interest in topical research.

The study is performed on the 25 countries forming European Union before the last wave of extension (Gligor and Ausloos, 2007).

The conclusions are drawn in the last section, where the predictive power and possibility of expanding some of the presented models are also discussed.

9.2 THE RANDOM CLUSTER MODEL FOR GROUP DECISION MAKING (GALAM, 2000b)

Millions of people make everyday decisions about some issue of interest for the group or the community. The complex process of group decision making supports a very wide class of phenomena. It includes professional groups having to decide on technical issues, as well as all kinds of friendly groups. The nature of the decision itself is even more diversified. It can be a high rank political issue related to some military retaliation, a public jury to decide whether someone is guilty or not, but it can equally be casual with some school board to decide about the opportunity of one optional topic for pupils to study.

Of course, the above decisions are of a totally different nature. The main differences are related to the associated human consequences and the decision costs. The main working hypothesis in sociophysics modeling is the existence of some universal features that are independent of both the social nature of the individuals making the decision and the nature of the decision itself.

The model might explain why all over the world and more specifically in democratic countries public opinion seems to be rather conservative. Even when the changes are desperately needed, an initial hostile minority appears to be always able to turn majority along its refusal position, preserving in this way the present state.

We consider a population with N individuals, which have to decide whether or not to accept a reform proposal. At time t, the proposal has a support by $N_+(t)$ persons leaving $N_-(t)$ individuals against it. Each person is supposed to have an opinion, i.e., $N_+(t) + N_-(t) = N$. The associated individual probabilities are thus $P_\pm(t) = N_\pm(t)/N$, with $P_+(t) + P_-(t) = 1$.

From this initial configuration, people start discussing the project. They do not meet all the time and all together at once. The random geometry of social life (within physical spaces like offices, houses, and schools) determines the number of people, which meet at a given place. Usually it is of the order of just a few (friends, acquaintances, and colleagues). The groups ("clusters") may be larger but it is rare. Since discussions can occur by chance, a given social life yields a random local geometry landscape characterized by a probability distribution for cluster sizes $\{m_i\}$ which satisfy the constraint $\sum_{i=1}^{M} m_i = 1$, where $i = 1$, $2,\ldots$, M denotes respective sizes $1, \ldots, M$, with M being the larger group.

People gatherings occur in sequences in time, each one allowing a new local discussion. Thus, each person may change his/her mind with respect to the reform proposal. During these meetings, all individuals are assumed to be involved in one group gathering. It means a given person is, on average, taking part to a group of size i with probability m_i. This working assumption becomes more realistic accepting the existence of one-person groups. Each new cycle of multisize discussions is marked by a time increment $+1$.

Note that no advantage is given to the minority with neither lobbying nor organized strategy. Moreover, an identical individual persuasive power is assumed for both sizes, and only the local majority argument determines the outcome of the discussion. People align along the local initial majority view. A group with an even number of members might be in the situation of fifty-fifty percent votes. Such a group is then within a nondecisional state. Here an equivalent of "inertia principle" from physics may be applied, in relation to the fundamental psychological asymmetry between what is known and what is hypothetical. *To go along with what is unknown, a local majority of at least one person is necessary.* Contrary, the full group turns against the reform proposal to preserve the existing situation.

Accordingly above considerations, having $P_{\pm}(t)$ at time t, at the time $(t+1)$, one gets:

$$P_+(t+1) = \sum_{m=1}^{M} a_m \sum_{j=[m/2+1]}^{m} C_j^m P_+(t)^j P_-(t)^{m-j} \qquad (9.1)$$

where $C_j^m = m!/(m-j)!j!$ and $[m/2+1]$ signifies the integer part of $(m/2+1)$.

Simultaneously

$$P_-(t+1) = \sum_{m=1}^{M} a_m \sum_{j=[m/2]}^{m} C_j^m P_-(t)^j P_+(t)^{m-j} \qquad (9.2)$$

In the course of time, the same people will meet again randomly in the same cluster configuration. At each new encounter, they discuss locally the issue at stake and change their mind according to the above majority rule. Repeated successive local discussions drive the whole population to a full polarization. Accordingly, public opinion *is not volatile*. It stabilizes rather quickly (after a small number of temporal steps) to a clear stand.

In Figure 9.1, one can see the variation of $P_+(t)$ as function of t (the temporal step). The initial value at is $P_+(0) = 0.60$. We considered a population of $N = 50$ individuals, whose decision (" +1"≡"Yes"; " −1"≡"No") are randomly distributed by a Monte Carlo algorithm for generating random numbers from a certain range (Mosetti et al., 2006). We fixed the values of M, but the same algorithm establishes the size and the position of clusters. The three time series correspond to $M = 2$, $M = 3$, and $M = 6$. To make a quantitative illustration of

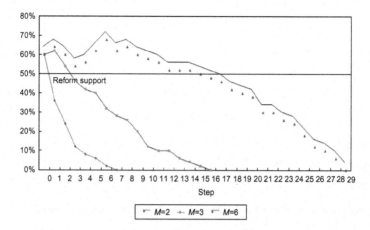

Figure 9.1 The dynamics of reform support, P₊, starting from a majority of 60%. Series 1: M = 2; *Series 2:* M = 3; *Series 3:* M = 6. *The steps of simulation may be considered as days of public debates.*

the dynamics refusal, let us consider the second case. The associated time series is: $P_+ = 0.60$; 0.62; 0.54; 0.46; 0.42; 0.40; 0.32; 0.28; 0.26; 0.20; 0.12; 0.10; 0.10; 0.06; 0.04; 0.02; 0.00 (Figure 9.1).

Sixteen cycles of discussion make all 60% of reform supporters to turn against it by merging with the initial 40% of reform opponents. Considering a discussion cycle corresponding to 1 day (in average), less than 3 weeks is enough to a total crystallization of opinion against the reform proposal. Note that a majority against the reform is obtained already within 3 days.

9.3 THE CLUSTER FORMATION IN THE SOCIAL IMPACT MODEL (KACPERSKI AND HOLYST, 2000)

The system consists of $N = 40$ individuals (members of a social group). Each of them can share one of two opposite opinions on a certain subject, denoted as $f_i = \pm 1$, $i = 1, 2, \ldots, N$.

Individuals can influence each other, and each of them is characterized by the parameter $y_i > 0$ which describes the strength of his/her influence. Every pair of individuals (i, j) is ascribed a distance d_{ij} in a social space. The changes of opinion are determined by the social impact exerted on every individual:

$$I_i = -y_i \beta - f_i h - \sum_{j=1, j \neq i}^{N} \frac{y_j f_i f_j}{g(d_{ij})} \qquad (9.3)$$

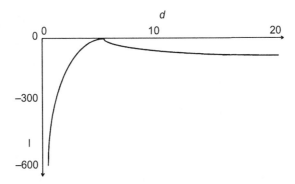

Figure 9.2 Cluster's radius a as a function of the leader's strength s_L for the circular social space.

where $g(x)$ is an increasing function of social distance. β is a self-support parameter reflecting the inclination of an individual to maintain his/her current opinion; h is an additional (external) influence which may be regarded as a global preference toward one of the opinions stimulated by mass media, government policy, etc.

The opinions of the individuals may change simultaneously in discrete time steps according to the rule:

$$f_i(t+1) = f_i(t) \tag{9.4a}$$

with the probability

$$\frac{\exp(-I_i/T)}{\exp(-I_i/T) + \exp(I_i/T)}$$

$$f_i(t+1) = -f_i(t) \tag{9.4b}$$

with the probability

$$\frac{\exp(I_i/T)}{\exp(-I_i/T) + \exp(I_i/T)}$$

The parameter T may be interpreted as a "social temperature" describing a degree of randomness in the behavior of individuals, while the impact I is a "deterministic" force inclining the individual i to change his/her opinion if $I_i > 0$, or to keep it otherwise (Holyst et al., 2000; Kacperski and Holyst, 2000). For example, in Figure 9.2 we have plotted the cluster's radius a as a function of the leader's strength s_L for the circular social space.

The simulation is performed assuming that the social space is a two-dimensional disk of radius $R \gg 1$, with the individuals located on the nodes of a quadratic grid. The distance between the nearest neighbors equals 1. The strength parameters s_i of the individuals are positive random numbers. In the center of the disk, there is a strong individual (the "leader"); his/her strength y_L is much larger than that of all the others ($y_L \gg y_i$).

At $T = 0$, the dynamical rule (9.4) becomes strictly deterministic:

$$f_i(t + 1) = -\text{sign}(I_i f_i) \tag{9.5}$$

Considering the possible stationary states, we find either the trivial unification (with equal opinion ± 1 for each individual) or, due to the symmetry, a circular cluster of individuals who share the opinion of the leader. This cluster is surrounded by a ring of their opponents (the majority). These states remain stationary also for a small self-support parameter (β), while for β sufficiently large, any configuration may remain "frozen."

Using the approximation of continuous distribution of individuals (i.e., replacing the sum in (9.3) by an integral), one can calculate the size of the cluster, i.e., its radius a as a function of the other parameters (Kacperski and Holyst, 2000).

9.4 FROM CLUSTERS OF INDIVIDUALS TO CLUSTERS OF COUNTRIES (GLIGOR AND AUSLOOS, 2007)

9.4.1 Mapping the Macroeconomic Time Series in Weighted Networks

In the 1780s, Euler invented network theory (Euler, 1736) for the Königsberg Bridge crossing problem, but this subject remained a form of abstract mathematics. Nowadays, many applications of the graph theory (Biggs et al., 1976) exist under the form of network analysis.

An increasing interest in the topics can be registered during the last decade, particularly due to its potentially unbounded area of applications. Indeed, the transdisciplinary concept of "network" is frequently met in all scientific research areas, its covering field spanning from computer science to medicine, economics, and social psychology. Moreover, it proves to be a reliable bridge between the natural and social sciences, so that the recent interest in its application to econophysics and sociophysics is also fully justified.

Using the strong methodological arsenal of mathematical graph theory, one has mainly focused on the dynamical evolution of networks, i.e., on the statistical physics of growing networks. The remarkable extension from the concept of classical random graph to the one of nonequilibrium growing networks allows for accounting the interest on the structural properties of random/complex networks in communications, biology, social sciences, and economy (Newman, 2003). Indeed, the field of possible applications seems to be unbounded, spanning from the World Wide Web and Internet structures to some more sophisticated social networks of scientific collaborations, paper citations, or collective listening habits and music genres (Dorogovtsev and Mendes, 2003).

In most approaches, the Euler graph theory legacy was preserved, especially as regards to the "Boolean" character of links: two vertices can be either tied or not tied, thus the elements of the so-called adjacency matrix only consist of zeros and ones. However, many biological and social networks, and particularly almost all economic networks, must be characterized by different strengths of the links between vertices. This aspect led to the concept of "weighted network" as a natural generalization of the graph-like approaches. Of course, various ways of attaching some weights to the edges of a fully connected network have been proposed in the recent literature (Newman, 2004). On the other hand, it is of major interest in economy to extract as much information as possible from the sparse and noisy macroeconomic (ME) time series. One must have in mind that most ME indicators have a yearly or at most quarterly frequency. When an ME indicator time series has been produced for a very long period, strong evidence against stationarity often arises. The macroeconomic indicator that we choose to investigate is the annual growth rates of GDP/capita. (Ausloos and Gligor, 2008; Gligor and Ausloos, 2007, 2008a, 2008b). Indeed, the GDP/capita is expected to reflect to the largest extent what Smith called over two centuries ago, "the wealth of nations." In fact, it is expected to account for both economic development and the well-being of the population. The target group of countries is composed of $M = 25$ countries: the 15 members of the European Union in 2004 (EU-15) and the 10 countries which joined the European community in 2005 (EU-10). Given the target country group, the World Bank database is used. In this way, the investigated time span goes from 1990 to 2005.

The correlation coefficients C_{ij} between two ME time series $\{x_i\}$ and $\{y_j\}$, $i, j = 1, \ldots, N$, are calculated, in this framework, according to the (Pearson's) classical formula. Each C_{ij} is clearly a function of both the time window size T and the initial time (i.e., the "position" of the constant size time window on the scanned time interval).

Let us now consider that the M agents (countries) which the ME time series refer to may be the vertices of a weighted network. The weight of the connection between i and j reflects the strength of correlations between the two agents and can be simply expressed as

$$w_{ij}(T) = |C_{ij}(T)| \qquad (9.6)$$

fulfilling the obvious relations $0 \leq w_{ij} \leq 1$; $w_{ij} = w_{ji}$ and $w_{ij} = 1$ for $i = j$.

One must stress at this point that the link connecting the vertices i and j does not reflect here either underlying interaction. Instead, the weight w_{ij} is a measure of the similarity degree between the ME fluctuations in the two countries. The term "fluctuations" refers here to the account of the annual rates of growth of the considered ME indicator. Networks are characterized by various parameters. For instance, the vertex degree is the total number of vertex connections. It may be generalized in a weighted network as

$$k_i = \sum_{\substack{j=1 \\ j \neq i}}^{M} w_{ij} \qquad (9.7)$$

Thus, the average degree in the network is

$$\langle k \rangle = \frac{1}{M} \sum_{i=1}^{M} \sum_{\substack{j=1 \\ j \neq i}}^{M} w_{ij} \qquad (9.8)$$

9.4.2 The Clustered Weighted Network of EU-25 Countries

A general question facing researchers in many areas of inquiry is how to organize observed data into meaningful structures. Having computed the adjacency matrix entries w_{ij}, some statistical properties of the $\{w_{ij}\}$ dataset are analyzed.

From the eigenvectors corresponding to the $[w_{ij}]$ matrix eigenvalues, a cluster-like structure can be built on the basis of the eigenvector components. Since the eigenvectors corresponding to the largest eigenvalues of the correlation matrix are usually expected to be those carrying the most useful information, a cluster-like structure of the EU-25 countries is built in Figure 9.3 on the basis of the structure of the first two eigenvectors.

One can easily see (Figure 9.3) that a multipolar structure exists: the "Continental" group (LHS, up) and the "Scandinavian" group (LHS, middle) are somewhat apart from each other. An extreme position is taken by GBR (RHS, down) which appears as the single member of any "Anglo" pattern, since the other OECD representatives of a (supposed to be) Anglo-convergence club, e.g., United States, Canada, and Australia), are missing from our study. Another interesting aspect is also found here, i.e., IRL has a nonaffiliation to the "Anglo" cluster, but is rather in the "Scandinavian" group and close to the "Continental" one.

One can also observe an emerging East European convergence club (RHS, down), tying to Scandinavian and Continental group

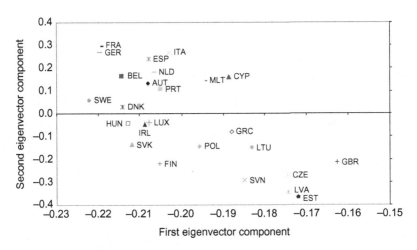

Figure 9.3 The cluster-like structure of the EU-25 countries according to the GDP/capita rates of growth. The country coordinates are the corresponding eigenvector components of the EU-25 weighted network adjacency matrix [w$_{ij}$].

through a HUN−POL line. Some "Mediterranean" clustering can be noticed as well.

The clustering scheme in Figure 9.3 is in agreement with results reported in the recent economic literature as regards the so-called convergence clubs across the Western Europe, i.e., groups of economies that present a homogeneous pattern and converge toward a common steady state (Aaberge et al., 2002; Angelini and Farina, 2005; Mora, 2005). In particular, in (Aaberge et al., 2002), it has been showed that Sweden, Norway, and Denmark registered a similar level of income mobility, while in Mora (2005), three distinct patterns of development and income distribution, indeed called "Continental," "Anglo," and "Scandinavian," have been found by examining a group of 17 OECD economies during the two decades before 2000. In the same idea, four clusters of different systems of social protection of OECD countries namely Scandinavian, Continental, Anglo-Saxon, and Mediterranean have been reported in Angelini and Farina (2005).

Finally, the adjacency matrix $[w_{ij}]$ can be used to plot the EU-25 network along a more geographical perspective. The network is, obviously, fully connected; it is of interest to observe the relative importance of the link strength (weights) through a display at different threshold values.

In the subsequent figures, only the links having the corresponding weights greater than a certain threshold value are taken into account. This threshold value is *a priori* chosen according to a significance level derived from the Student's *t*-statistic test applied to the $[w_{ij}]$ matrix (Gligor and Ausloos, 2007). The resulting networks are plotted in Figures 9.5 and 9.6. The reduced network (Figure 9.5) includes all four "convergence clubs" above discussed. The single element "Anglo" club is well isolated, as seen already in Figure 9.3; the "Scandinavian" and "East European" clusters become isolated for a higher threshold, as seen in the network (Figure 9.6).

It is also remarkable the decreasing number of long-range links when going from Figure 9.4 to Figure 9.5 and further to Figure 9.6. Even if the actual geographic, investment, and trade inter-country ties are not explicitly considered in our study, the degree of similarity of the country GDP/capita fluctuations well supports the evidence of the so-called regionalization.

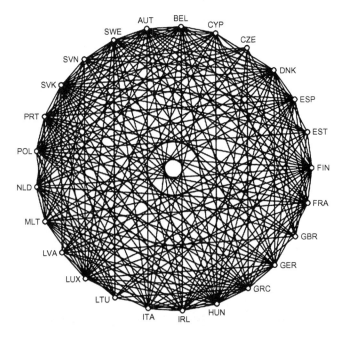

Figure 9.4 The EU-25 initial weighted network. The threshold of weights: $w_T = 0$.

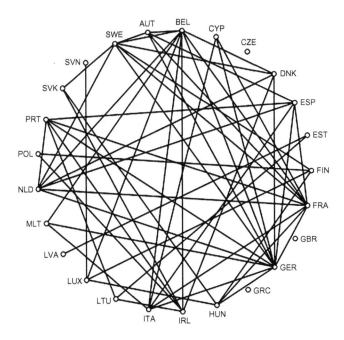

Figure 9.5 The EU-25 weighted network includes all four "convergence clubs." The threshold of weights: $w_T = 0.69$.

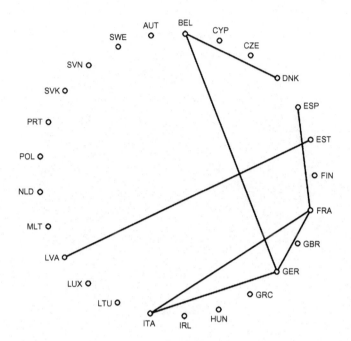

Figure 9.6 The EU-25 weighted network where the "Scandinavian" and "East European" clusters become isolated. The threshold of weights: $w_T = 0.81$.

9.4.3 The Country Overlapping Hierarchy

The previous results lead to consideration of the hierarchical or "clustering" structure of the EU-25 countries. For the purpose of describing this aspect, we introduce a quantity that is able to measure to what extent a country is "connected" to the whole system. The idea, first hereby applied to a nonweighted network, is to construct a country hierarchy using some new coefficient, O_{ij}, which takes into account not only the degrees k_i and k_j, but also the number N_{ij} of the common neighbors of i and j vertices. This coefficient O_{ij} is here called "overlapping" of i and j vertices (in spite of the fact that this term has already been assigned various meanings in the network literature).

Firstly, for a classical random network consisting of M vertices, the overlapping coefficient O_{ij} must satisfy the following properties:

1. $O_{ij} = 0 \Leftrightarrow N_{ij} = 0$ (fully disconnected, or "tree-like" network).
2. $O_{ij} = 1$, $\forall i \neq j$ in a fully connected network, where
   $$N_{ij} = M-2; \ k_i = k_j = M-1;$$
3. $0 < O_{ij} < 1$, otherwise.
4. $O_{ij} \sim N_{ij}$ and $O_{ij} \sim \langle k_{ij} \rangle \equiv (k_i + k_j)/2$.

A quantity satisfying all these conditions (1)–(4) can be defined as

$$O_{ij} = \frac{N_{ij}(k_i + k_j)}{2(M-1)(M-2)}, \quad i \neq j \qquad (9.9)$$

For a weighted network, the above relation may be generalized as

$$O_{ij} = \frac{1}{4(M-1)(M-2)} A_{ij}, \quad i \neq j \qquad (9.10)$$

where

$$A_{ij} = \sum_{\substack{l=1 \\ l \neq i,j}}^{M} (w_{il} + w_{jl}) \left(\sum_{\substack{p=1 \\ p \neq i}}^{M} w_{ip} + \sum_{\substack{q=1 \\ q \neq j}}^{M} w_{jq} \right) \qquad (9.11)$$

One can easily see that $0 < O_{ij} < 1$ and $O_{ij} = 1$ only for all $w_{ij} = 1$, i.e., fully connected nonweighted network. However, for a weighted network, O_{ij} can never be zero.

Each overlapping coefficient is thus computed for each EU-25 country using the adjacency matrix defined in Eq. (9.11). A country average overlapping index $\langle O_i \rangle$ can be next assigned to each country, dividing the sum of its overlapping coefficients by the number of neighbors:

$$\langle O_i \rangle = \frac{1}{M-1} \sum_{j=1}^{M} O_{ij} \qquad (9.12)$$

The results are given in Table 9.1.

The highest values of the average overlapping index correspond to the countries belonging to the "Continental" and "Scandinavian" groups, while the lowest values correspond to several countries forming an East European cluster. Again, the separate position of GBR as a single representative of the "Anglo" pattern, with respect to European economies is emphasized, the former being in fact a cluster by itself.

One has to note a remarkable similarity between the country ranking over the first eigenvector component (Figure 9.3) and the ranking

Table 9.1 The Country Average Overlapping Index of Each EU-25 Country					
SWE	0.38	ESP	0.35	CYP	0.32
GER	0.37	AUT	0.35	SVN	0.32
FRA	0.37	FIN	0.35	LTU	0.32
DNK	0.37	PRT	0.35	LVA	0.31
HUN	0.37	NLD	0.35	CZE	0.31
SVK	0.37	ITA	0.35	EST	0.30
BEL	0.36	POL	0.34	GBR	0.29
IRL	0.36	MLT	0.33		
LUX	0.36	GRC	0.33		

over the country average overlapping index (Table 9.1). This similarity proves the ability of the hereby introduced $\langle O_i \rangle$ index to supply a correct description of the country weighted network.

9.5 CONCLUSIONS

Group decision making plays a central role in topical sociophysics. Section 9.2 presented several applications of the real space renormalization group techniques to the study of the dynamics of representatives elected using the local majority rule voting. It was shown that, on one hand, majority rule voting produces critical thresholds to full power, and, on the other hand, why an initial hostile minority appears to be able to turn the majority along its refusal position, preserving in this way the present state. Using models of this kind, several real political events were successfully predicted (Galam, 2000a, 2000b; Le Hir, 2005). Section 9.3 presented the opinion dynamics in the social impact models. The model captures the main aspect of the process of opinion formation, namely the cooperative phenomenon due to instantaneous mutual influence of individuals. The leader's influence and the noise effect were also studied, by applying some nonlinear dynamics methods. The model supplies a tool for description of an important mechanism inducing changes of attitudes, which is a cooperative phenomenon difficult to explain in the traditional descriptive sociological language.

In Section 9.4, we pointed out that mapping the GDP/capita and other ME time series into a weighted network structure allows a direct visualization of the inter-country connections from at least three different viewpoints: (a) as relative distances in the bi- or multidimensional space of the adjacency matrix eigenvalues; (b) as statistical significant edges in the graph plot; (c) as relative positions in the country averaged overlapping coefficients-based hierarchy. In all these three ways, the derived clustering structure is found to have a remarkable consistency with the results reported in the actual economic literature. Thus, other network multivertex characteristics (clustering, minimal path, centrality, etc.) may also be expected to play an important role in a better understanding of social–economic connections.

REFERENCES

Aaberge, R., Bjorklund, A., Jantti, M., Palme, M., Pedersen, P.J., Smith, N., Wennemo, T., 2002. Income inequality and income mobility in the Scandinavian countries compared to the United States. Rev. Income Wealth 48, 443–469.

Angelini, E.C., Farina, F., 2005. The size of redistribution in OECD countries: Does it influence wave inequality? Paper for International Conference in Memory of Two Eminent Social Scientists: C. Gini and M. O. Lorenz, Siena, Italy, 23–26 May.

Ausloos, M., Gligor, M., 2008. Cluster expansion method for evolving networks having vector-like nodes. Acta Phys. Pol. A 114 (3), 491–499.

Biggs, N.L., Lloyd, E.K., Wilson, R.J., 1976. Graph Theory. Oxford University Press, Oxford.

Bouchaud, J.P., Mézard, M., 2000. Wealth condensation in a simple model of economy. http://xxx.lanl.gov/abs/cond-mat/0002374 (last accessed 02.10.2012).

Challet, D., Zhang, Y.-C., 1998. On the minority game: analytical and numerical studies. Phys. A 256 (3–4), 514–532.

Dorogovtsev, S.N., Mendes, J.F.F., 2003. Evolution of Networks: From Biological Nets to the Internet and WWW. Oxford University Press, Oxford.

Drăgulescu, A., Yakovenko, V.M., 2000. Statistical mechanics of money. Eur. Phys. J. B 17, 723–729.

Euler, L., 1736. Solutio problematis ad geometriam situs pertinentis. Commetarii Academiae Scientiarum Imperialis Petropolitanae 8, 128–140.

Galam, S., 1986. Majority rule, hierarchical structure and democratic totalitarism. J. Math. Psychol. 30, 426–434.

Galam, S., 2000a. Les réformes sont-elles impossibles?. Le Monde 28 (Mars), 18–19.

Galam, S., 2000b. Real space renormalization group and totalitarian paradox of majority rule voting. Phys. A 285, 66–76.

Galam, S., 2002. Minority opinion spreading in random geometry. Eur. Phys. J. B 25, 403–406.

Giardina, I; Bouchaud, J.P., Mézard, M., 2000. Population dynamics in a random environment. http://xxx.lanl.gov/cond-mat/0005187 (last accessed 02.10.2012).

Gligor, M., 2001. Noise induced transitions in some socio-economic systems. Complexity 6, 28–32.

Gligor, M., Ausloos, M., 2007. Cluster structure of EU-15 countries derived from the correlation matrix analysis of ME index fluctuations. Eur. Phys. J. B 57, 139–146.

Gligor, M., Ausloos, M., 2008a. Clusters in weighted ME networks: the EU case. Eur. Phys. J. B 63, 533–539.

Gligor, M., Ausloos, M., 2008b. Convergence and cluster structures in EU area according to fluctuations in macroeconomic area. J. Econ. Integr. 23, 297–330.

Gligor, M., Ignat, M., 2001. Some demographic crashes seen as phase transitions. Phys. A 301, 535–544.

Gligor, M., Ignat, M., 2003. A random cluster model for group decision-making. The Scientific Review "V. Adamachi" XI (1), 3–4. Ed. University "Al. I. Cuza", Iasi.

Guzmán-Vargas, L., Hernández-Pérez, R., 2006. Small-world topology and memory effects on decision time in opinion dynamics. Phys. A 372 (2), 326–332.

Haken, H., 1983. Advanced Synergetics. Springer, Heidelberg.

Holyst, J.A., Kacperski, K., Schweitzer, F., 2000. Phase transitions in social impact models of opinion formation. Phys. A 285, 199–210.

Kacperski, K., Holyst, J.A., 2000. Phase transitions as a persistent feature of groups with leaders in models of opinion formation. Phys. A 287, 631–643.

Laguna, M.F., Abramson, G., Zanette, D.H., 2003. Vector opinion dynamics in a model for social influence. Phys. A 329 (3-4), 459–472.

Le Hir, P., 2005. Les mathématiques s'invitent dans le débat européen. Le Monde 26 (Février), 23–24.

Malescio, G., Dokholyan, N.V., Buldyrev, S.V., Stanley, H.E., 2000. Hierarchical organization of cities and nations. http://xxx.lanl.gov/abs/cond-mat/0005178 (last accessed 02.10.2012).

Montroll, E.W., 1987. On the dynamics and evolution of some socio-technical systems. Mon. Bull. Am. Math. Soc. 16, 1–46.

Mora, T., 2005. Evidencing European regional convergence clubs with optimal grouping criteria. Appl. Econ. Lett. 12, 937–940.

Mosetti, G., Challet, D., Zhang, Y-C., 2006. Minority games with heterogeneous timescales. Phys. A 365 (2), 529–542.

Newman, M.E.J., 2003. The structure and function of complex networks. SIAM Rev. 45, 167–256.

Newman, M.E.J., 2004. Analysis of weighted networks. Phys. Rev. E 70, 056131.

Stauffer, D., Sahimi, M., 2007. Can a few fanatics influence the opinion of a large segment of a society? Eur. Phys. J. B 57, 147–152.

Tryon, R.C., 1939. Cluster Analysis. Editure: Edwards Brothers, Ann Arbor, MI.

CHAPTER 10

Sociophysics: A New Science or a New Domain for Physicists in a Modern University

Gheorghe Săvoiu and Ion Iorga Simăn
University of Piteşti, Faculty of Economics and Faculty of Sciences, Romania

10.1 INTRODUCTION

A new attitude comes into focus ever more clearly in the process that tends to separate sociophysics from the broader area of econophysics and especially through its increasingly obvious independent development.

Sociophysics as a new science has developed concepts, including standardized typologies, definitions, and measures of key concepts and consensus statements, techniques, tools, and methods, and has tested new specific theoretical models and conceptual frameworks to address contemporary sociological challenges, to capture systematic information from social domains and to develop an implementation context using physics-related thinking. Scientific research and the practice of sociophysics are defining a new era with a more intense use of physical models from statistical physics or with a wider view of data in the specific way of thinking of quantum statistics. In 1902, when Josiah Willard Gibbs published *Elementary Principles in Statistical Mechanics* at Yale's Publishing House, the father of the newly born science simply called it 'statistical mechanics' he certainly did not know or did not imagine that this new inter-, trans-, and multidisciplinary science would be so relevant for the study of nonphysical systems. After more than 100 years, methods and models of statistical mechanics or statistical physics can be successfully applied to social problems. The vast experience of physicists in working with experimental data gives them certain advantage to uncover quantitative laws in the statistical data available in sociology. Sociophysics brings new insights and new perspectives, which are likely to revolutionize the old social disciplines. If some agreements are possible between the economists or sociologists and the physicists, it is probably about the need for a more intense exchange of information. Statistical physics as the first method for sociophysics has proven to be a very fruitful framework for describing phenomena outside the realm of traditional physics. The last years have witnessed the attempt by physicists to study collective phenomena emerging from the interactions of individuals as elementary units in population and social structures.

This chapter is organized as a review of the qualitative improvement brought about by sociophysics. The following section deals with a brief history and some activities and models, the next section concentrates on some definitional issues for this new science, is more focused on contemporary and future trends and activities in sociophysics.

10.2 A BRIEF HISTORY OF SOCIOPHYSICS

In 1835, the publication of *Physique Sociale* by Adolphe Quetelet, a pioneering book containing an original view of social statistics, was the first scientific approach to the social sciences through new mathematical methods. For a couple of hundred years, statistics was treated as a social discipline (only in the last 100 years was it considered in mathematical terms, a special universal science about mass phenomena, regardless of its nature). Even now it is irreplaceable in terms of social sciences statistics, whereas for other natural sciences, it is just one of the tools. But cohabitation of physics and social sciences is a much more complicated phenomenon because the objects are not physical any longer (people instead of particles, human interactions instead of molecule collisions, etc.). Physics was used in social studies only as an analogy, becoming only a methodological tool, and in that capacity it openly competes with already existing and very well-developed statistics. But certainly physics may indeed change traditional "overstatisticized" view of society and enrich it (Chakraborti, 2008).

Sociophysics has become an attractive field of research over the last two or three decades, despite the controversies between sociologists and sociophysicists and its potential use for understanding the social phenomena.

Sociophysics aims at a statistical physics modeling of large-scale social phenomena, like culture and opinion formation and dynamics, cultural and behavioral dissemination, the origin and evolution of language, competition and conflicts, crowd behavior, social contagion, gossip and rumors evolutions, Internet and World Wide Web, cooperation and scientific research, and occurrences of terrorism. Sociophysics tries to model the dynamics of social and economic indicators of a society and investigate how the introduction of life extension will influence fertility rates, population growth, and the distribution of wealth (Mantegna and Stanley, 2000), religion, ecosystems, friendship and sex, social network and traffic, too.

After more than 100 years, the methods and techniques of statistical physics can be successfully applied not only to economic but also to social problems. "Today physicists regard the application of statistical mechanics to social phenomena as a new and risky venture. Few, it seems, recall how the process originated the other way around, in the days when physical science and social science were the twin siblings of a mechanistic philosophy and when it was not in the least disreputable to invoke the habits of people to explain the habits of inanimate particles" (Ball, 2004).

The origins of modern sociophysics are traced back in its history to the late 1970s and 1980s. One of the leading authors in sociophysics, Serge Galam, published his early works in the *Journal of Mathematical Sociology* (Galam et al., 1982), *Journal of Mathematical Psychology* (Galam, 1986), *Journal of Statistical Physics* (Galam, 1990), *International Journal of General Systems* (Galam, 1991), and the *European Journal of Social Psychology* (Galam and Moscovici, 1991), among others. The conflicting nature of sociophysics with the physics community was revealed by Serge Galam's account of his experience in 'Sociophysics: a personal testimony' (2004).

Physics has influenced the social sciences since the times of Galileo and Newton. The ideas of Schumpeter about the influence of innovations on society are important proofs of the last observation. After one century of understanding the relativity theory and 80 years after the establishing of quantum mechanics, physics turns to new areas of the complex systems research. Up to the last two or three decades, these regions of research have been reserved for sociology. The first interest of physicists in social sciences systems has roots that date back to 1936, when Majorana wrote a pioneering paper, published in 1942 and entitled *Il valore delle leggi statistiche nella fisica e nelle scienze sociali*, on the essential analogy between statistical laws in physics and social sciences. Many years later, statistical physicist Elliott Montroll coauthored with W.W. Badger, in 1974, the book entitled *Introduction to Quantitative Aspects of Social Phenomena*. Physics is now concentrated on scientific and technological aspects of human society and accepts the ideas of Alfred Lotka on populations as energy transformers, the dynamics of technical invention capacity of the society, or the population dynamics models detailing hypotheses about migration between two geographic regions, etc. Physics emphasizes the need for more investigation into the social processes by means of the modern methods of mathematics, statistics, and sociology in the new science of sociophysics, that aims at a statistical physics modeling of large-scale social phenomena, like opinion formation, cultural dissemination, the origin and evolution of language, crowd behavior, social contagion, and interactions of individuals as elementary units in social structures. A lot of work in new sociophysics has been carried on, especially in the design of microscopic models, whereas comparatively little attention has been paid to a quantitative description of social phenomena and to the promotion of an effective cooperation between physicists and social scientists. The name 'sociophysics' has been around for decades, but only in the twenty-first century did it become more of a science than a slogan.

Sociophysics is a much less studied and published topic than econophysics, another new border science or new domain for physics. Initially named psychophysics, sociophysics can be described as the sum of activities of searching for fundamental laws and principles that characterize human behavior and result in collective social phenomena. In this domain of econophysics are included topics such as the dynamics of complex social networks (which is how the above work ties in here), robustness of social processes, the scaling of social systems, and the evolution of social organization. Each of these subtopics represents a union of what it is called a sociophysics perspective with approaches from other fields. An anecdotic effect, like the movement of a butterfly's wings that can affect the weather, is an example of the sensibility of the physical system behavior with respect to its initial conditions, a fingerprint of the deterministic chaos. But it is also an example of how much importance even the smallest detail can assume within sociophysics. The gap between empirical sociology and modern sociophysics is perhaps smaller than all the others occurred between the hermeneutic and humanistic social sciences.

10.3 DEFINITIONAL ISSUES OF SOCIOPHYSICS

The study of behavioral and social phenomena has experienced a surge of interest over the last decade. One reason for this great attention paid to sociophysics is the huge amount of high-quality data made available by Internet technologies. Also much of modern sociological research falls under the umbrella of sociophysics and brings a physics perspective to the problem of complex collective behaviors. Thus, the apparently common field of sociology has the potential of producing and proving that the laws of physics can be reproduced as laws in human interaction, in social constructions, and even in relationships.

In sociophysics, the first objective is the treatment of individuals, somewhat analogously to particles, or to atoms in a gas, and this allows for the application of statistical physics methodologies. Sociophysics describes a lot of the aspects of social and political behavior, considering that human individuals behave much like atoms concepts from the physics of disordered matter (Galam, 2004; 2008a, 2008b; 2011). Why not to do it, if this new science called sociophysics allows people to achieve more social freedom in the real world, and offers sociologists a better comprehending of the richness and potential of our social and political interactions? The second objective of sociophysics is, of course, to offer the intuitive/psychic information using physics' models and methods, which emerges simultaneously and complements the theoretical applications even in politics (e.g., Paris

Arnopoulos wrote a famous book, *Sociophysics,* in 1993, in which he speculates on heat, pressure, temperature, entropy, and volumes of societies, and Serge Galam and Snajd offered their famous sociophysical logical models during the last decades). Sociophysics managed to open a new path to a radically different vision of society and personal responsibility (Vitanov et al., 2005; Vitanov et al., 2010).

Psychic information and intuitive guidance are tools that are both a natural right for us and can also be used to improve our mind integration here on Earth. Psychic information can be accessed by anyone and everyone; however, sometimes it is harder for us to hear and receive information about our own life, and that is where sociophysics comes in, like a "deus ex machina," but not such as an "angel" suddenly appearing to solve complex social problems but in a scientific and methodological manner. Two particular methods of this theory are applied in sociophysics relatively often: the master equations, an analytical, relatively easy and approximate method (Helbing, 1995; Helbing, 2010; Weidlich, 2000), and the Monte Carlo simulation, a numerical, technically difficult and, in principle, exact method (De Oliveira et al., 1999). Various and numerous social processes were attempted to be described with these methods: migration dynamics, residential segregation, competitions, gossip, evolution of cultures and languages, opinion dynamics, and many others (Klemm et al., 2005; Lind et al., 2007; Schulze and Stauffer, 2006; Sobkowicz, 2011; Stauffer, 2003, 2005; Stauffer et al., 2006; Sznajd-Weron, 2005; Sznajd-Weron and Sznajd, 2000; Sznajd-Weron and Pekalski, 2001; Sznajd-Weron, 2004; Bernardes et al., 2001; Hu et al., 2012).

Sociophysics' method applied on new ideas is multidirectional, sequential, complex, and original. Thus, the references to complexity, diffusion, entropy, self-organization, randomness, fluctuations, fractals, criticality, and chaos can be found in papers on sociophysics. Barkley J. Rosser, Jr, has identified 12 new domains of sociophysics covered by more than 210 remarkable papers: culture (music, paintings, books), competition and conflicts, cooperation and scientific research, ecosystems, friendship and sex, Internet and World Wide Web, languages, opinion dynamics, power laws and fractals, religion, social networks, and traffic. As good methods flow to other areas, scientific researchers immediately declared a new kind of science, as a consequence of the result of unified knowledge and of the interdisciplinary field applications. If there is a need to follow arguments put forth by Rosario Mantegna and Eugene Stanley in econophysics, what is involved in the definition of sociophysics is the phenomenon of physicists using their models to study sociology, which is itself a slightly curious way to define a scientific discipline, given that this is itself a functional and sociological definition (a physicist is doing something in a new domain like sociology) rather than one based on the content of the ideas contained in the new science's object (Rosser, 2006). First sociophysics was a new insight into the applicability of much of elementary statistical physics to the social sciences (Galam, 1984), but now it is much more than this, which means a new insight followed by transferring and further developing ideas and concepts common to physics, biology, and ecological systems.

In 2005, Igor Mandel and Dmitri Kuznetsov have introduced *Mediaphysics* as a part of sociophysics, studying processes of mass communications in social systems and demonstrated its potential for applications in different processes of mass communications in complicated social or sociobiological systems such as marketing, economics, politics, and animal populations.

Philip Ball's definition of sociophysics describes the new science as mostly simulations in which independent entities (e.g., particles, people, and institutions) act and react according to specific rules or laws (Ball, 2004). Another simple and clear definition of sociophysics underlines that it brings a physics perspective to the problem of complex collective behaviors. A wide variety of specific concepts are covered and a wide variety of specific methods are used in the new discipline called sociophysics. Perhaps, the entire field of sociophysics is nothing else but the unification of sociology and physics and studies how cause and effect, energy, magnetism, and human relationships meld, although with much originality and ingenuity. Gradually, sociophysics becomes a new and specialized discipline, which is also a system of thought and is reflected in its new methodological approach to the social phenomenon. A good overview of several fields of application and an accessible, entry-level description of many simulation models can be interpreted as forming part of sociophysics. For instance, in a paroxysmal crisis of fear, opinions can be activated very quickly among millions of mobilized citizens, ready to act in the same direction, against the same enemy; but a lot of phenomena can be studied within the newly emerging field of sociophysics, in particular the dynamics of minority opinion spreading the rumor propagation (Deffuant, 2002; Galam, 1999, 2002, 2003, 2008a, 2008b; Sznajd-Weron, 2005).

Some of the most remarkable pioneers of sociophysics probably are Serge Galam (Sociophysics: a personal testimony), Dietrich Stauffer (Sociophysics simulations I: language competition), Paris Arnopoulos (*Sociophysics: Chaos and Cosmos in Nature and Culture*), Katarzyna Sznajd-Weron (Sznajd model and its applications), etc.

Sociophysics needs more clarity, especially when it envisions probability at the foundation of social theory. There is no contradiction between this new field of sociophysics and statistics. But, certainly, sociophysicists should be more careful when they are justifying their complex models. Sometimes this action of minds seems to be averaged out and finally removed by virtue of the law of large numbers (Galam and Jacobs, 2007).

In the last two decades, new interdisciplinary approaches to economics and social science have been developed by natural scientists. The problems of economic growth, distribution of wealth, and unemployment require a new understanding of markets and society. The dynamics of social systems has been introduced by W. Weidlich (1972), the term econophysics has been coined by H.E. Stanley (1992), and Sociophysics has been used by S. Galam in a similar manner.

10.4 CONTEMPORARY AND FUTURE TRENDS AND ACTIVITIES

The challenging and peculiar feature of sociophysics' models is their ability to reproduce, in some respects, real social systems. For a better understanding, there are detailed two models of spreading opinions within a human population. Serge Galam was the first to model the spread of opinions within a population and get an equation of the inertia of democratic systems against changes. Over the last 25 years, sociophysicists have introduced a series of sociophysics models. These could be divided into more than 20 different general classes, which deal respectively with the following:

• Opinion dynamics
• Decision making

- Competitions/conflicts, fragmentation versus coalitions
- Income or wealth spreading and concentration
- Residential segregation and migration dynamics
- Cultures and languages evolution
- Friendship and sex
- Internet and World Wide Web evolution
- Spread of religion
- Social networks dynamics
- Traffic dynamics
- Democratic voting in bottom-up hierarchical systems
- Spread of terrorism
- Generalized social interaction processes
- Spread of rumors and gossip in social networks
- Model of consensus formation
- Globalization versus polarization phenomena
- Political environment and democracy trend
- Simulation of inherited longevity
- Election and dynamics or distribution of political votes
- Simulations for biological aging and sexual reproduction, etc.

Using original models, several major real political social and religious events were successfully predicted (from the victory of the French extreme right party in the 2000 election to the fifty–fifty ballot count in several democratic countries like Germany or Italy). The models are really important tools for a reasonable perspective and make sociophysics a predictive solid field. There are many research models specific to sociophysics, with different dynamic rules but with the same macroscopic behavior (from coarsening to critical exponents).

The following could be a restricted typology of the sociophysics models, based on the hypothesis that human beings think and act or behave like particles which have no feelings and no free will.

1. Opinion dynamics (Deffuant, Galam, Sznajd-Weron, Minority/ Majority rule).
2. Cultural dynamics (Axelrod, Levine).
3. Language dynamics or evolution (Abrams-Strogatz, Minett-Wang), etc.

Sometimes models are philosophical instruments more than scientific. In the year 2000, Katarzyna Sznajd-Weron proposed a model of opinion formation, which was based on the trade union maxim

"United we Stand, Divided we Fall" (USDF), known as the Sznajd model (SM). The main characteristic of the model is that information flows only outward.

This apparently very simple model was aimed at describing global social phenomena (sociology) by local social interactions (described by social psychology) and answered the question on how opinions spread in a human society. Social opinion is of course the outcome of individual opinions, represented by many sociologists earlier, in their models Ising spins ("yes" or "no"). This was the first Sznajd-Weron model, and the new model introduced an original concept of spin dynamics (Sznajd-Weron, 2005):

a. "In each time step, a pair of spins S_i and S_{i+1} is chosen to change their nearest neighbors (nn), i.e., the spins S_{i-1} and S_{i+2}.
b. If $S_i = S_{i+1}$, then $S_{i-1} = S_i$ and $S_{i+2} = S_i$ (social validation).
c. If $S_i = -S_{i+1}$, then $S_{i-1} = S_{i+1}$ and $S_{i+2} = S_i$."

Motivated by the phenomenon known as social validation, SM (renamed by Dietrich Stauffer) has been modified and applied such as from marketing to finance and from language dynamics to politics. The model was also improved by Serge Galam from one-dimensional SM and transformed exactly into two dimensions: the one-dimensional rule being applied to each of the four chains of four spins each, centered about two horizontal and two vertical pairs. Thus Serge Galam has changed the entire model and constructed two-dimensional version of the so-called two-component model.

A great hope for the model of sociophysics is to show similar correspondence between simple interactions among entities (agents being the preferred sociophysical term) and complex behavior in the final aggregate. In 2005, for the first time, a highly improbable political vote outcome was predicted using a sociophysics model.

In the next years, both physicists, on one hand, and psychologists and sociologists, on the other, will try to design a basic course to teach the students the basic elements from physics and psychology or sociology. Some of the new areas of opportunity for sociophysics are as follows:

1. Indexphysics or the new construction of economic and social indices (from consumer price index or CPI to human development index or HDI) because of the accuracy of statistical physics in its methods and techniques.

2. New understanding of natural economic life through the method of physics.
3. Physics of distribution or physics analysis of wealth, political, or economical power, and resources to optimize the dimension of firms, institution, and other socioeconomic entities.
4. Convergence and divergence on the micro-market and the spectrum of evolution for the macro-market with the best results in lower transaction costs and more efficient strategies.
5. Mediaphysics, proposed as a concept of analyzing communicational phenomena in societies, briefly considered as the most possible way to bridge two different paradigms (mass media and physics). An example of using mediaphysics principles on marketing material for a large company was presented in the paper STATISTICAL AND PHYSICAL PARADIGMS (ECONOPHYSICS, SOCIOPHYSICS, MEDIAPHYSICS) by Igor Mandel and Dmitri Kuznetsov.
6. A new concept of opinion changing rate that transforms the usual approach to opinion consensus modeling into a synchronization problem.
7. Terrorism risk emerged as a quantitative modeling discipline after 9/11, terrorist modus operandi being a function of human behavior, and so requires special methods drawn from fields such as game theory, social psychology, and network analysis.
8. The availability of high-volume and high-quality records of data allows us to experience and exploit concepts and methods traditionally belonging to the areas of statistical physics and complexity, in the social sciences: urban textures, the World Wide Web, and firms are described in terms of random structures in high-dimensional representation.
9. City size, income, word frequency, and music genres are distributed according to power laws and evolve under the effect of spatial–temporal correlations.
10. Typical of physical systems with many interacting units.
11. The dynamics underlying social conflicts and competition.
12. The insurgent group formation and attacks in all modern wars.
13. Airways systems.
14. Opinion dynamics in a bounded confidence consensus model (from continuum opinion dynamics model of Krause and Hegselmann to Santo Fortunato, Vito Latora, Alessandro Pluchino, Andrea Rapisarda Model).

15. Opinion changing rate model (OCRM), a modified version of the Kuramoto model, one of the simplest models for synchronization in biological systems.
16. Interacting agent models used to study bifurcations in group dynamics.
17. Social networks and crowd dynamics in traffic.
18. Quantum economics based on social behavior and economic opinion dynamics.
19. Global warming: a social phenomena, etc.

Somehow, physicists are still divided: some are convinced it will produce new understanding of economic and social phenomena, and some are dubious. Physics still has a lot to teach both psychology and sociology and more other social sciences:

- The effort that is put into getting data about processes.
- The importance of developing new methods of measurement.
- The importance that is given to abstraction and thus evidence over models and concrete models over frameworks and paradigms.
- The willingness to develop modeling techniques when the existing ones become inadequate.
- The acceptance of evidence for choosing between competing theories.

Sociophysics, in a transdisciplinary perspective, has transited from recognized subfields such as statistical physics and quantum physics to new sciences not only because these sciences are powerful tools in statistical physics or quantum physics (as well worked out as linear regression in econometrics or in mathematical models and results) but also because of the new way of learning and understanding reality in a great diversity of methods, and so the older sciences like physics, psychology, and sociology are more and more credible.

10.5 WHY SOCIOPHYSICS IS NECESSARY IN A MODERN UNIVERSITY?

Apparently, this question is not so difficult. But, let us make this job easier, tieing together a model of thinking based on contemporary paradoxes of our learning process and institutions.

10.5.1 What Does the Modern Concept of University Mean?
First Answer: *A place full of intelligent people...*

Sometimes, there are even more than in other places like the institutions for scientific research, banks, political institutions, or entities.

This interesting truth can generate the first law or the first Paradox of Universities, called Karl Albrecht's Law:

"Intelligent people, when assembled into an organization, will tend toward collective stupidity." (Albrecht, 2003, p. 4).

Does this process happen in a modern university, too? Maybe yes or maybe not. However, it does happen frequently since it follows the law of *entropy*, which measures energy degradation in a natural system through increasing disorder.

Karl Albrecht explains the generation of synergy in a knowledge field, introducing the concept of *syntropy: as the coming together of people, ideas, resources, systems, and leadership in such a way as to fully capitalize on the possibilities of each* (Albrecht, 2003, p. 42).

Well then, what about syntropy? Would it denote the upgrading of organizational energy? Entropy would then show the natural tendency of people toward loose interaction and increased stupidity. The same syntropy would show the conscious, deliberate, and intelligent effort for organizational learning (Brătianu, 2007a).

Under these circumstances, what could be the best model to understand the entropy and syntropy realities?

This could be the first argument that proves that econophysics is virtually the most necessary science and discipline in a modern university.

Second Answer: A place where there is a continuous process of enlarging democratic access and reaccess of more and more people to higher education and a learning organization where there is an increasing academic excellence process, too. Is the modern *organization* called university a social invention indeed?

An organization represents a systematic arrangement of people brought together to accomplish some specific objectives, impossible to be realized by one single man (Robbins and DeCenzo, 2005). The concept of organization has no absolute meaning since an organization is only a tool for making people productive in working together. It has a relative meaning. Actually, this is reflected in the origins of the word organization, which derives from the Greek *organon*, meaning a tool or instrument. That means an organization is not an end in itself but an instrument conceived to perform some kind of goal-oriented processes.

How does a contemporary university succeed in organizing people?

The process of management is necessary in order to perform this process efficiently and effectively. Efficiency means doing tasks correctly so that products can be obtained with minimum of resources. Effectiveness means doing the right task (a linear or a nonlinear thinking, but surely it is a deterministic kind of thinking).

But contemporary managerial decisions made in conditions of uncertainty generate risks, even in a modern university. In order to identify, evaluate, and accept risks we need to develop new thinking models based on random events and accept the importance of random thinking.

This could be the second argument for econophysics' model, about which, I hope we shall think again if it is really the best model for prognosis.

Third Answer: *A place based on the modern learning processes, on the recently scientific theories and on the most useful discoveries...*

How could be solved Bratianu's paradox formulated as follows: *Although a university is an organization based on learning processes, it is not necessarily a learning organization?* Since *learning* is a fundamental process within any university, people may consider universities as being learning organizations. This is a major error, especially in the former socialist countries. The purpose of a modern university is to demonstrate that they are far away from being learning organizations due to some organizational learning barriers (Brătianu, 2007a, p. 376).

A learning organization is "an organization that is continually expanding its capacity to create its future. For such an organization (as university is), it is not enough merely to survive" (Senge, 1990, p. 14).

Adaptive learning should be only the first phase of a modern university contemporary process, being continued with *generative learning*, the process that enhances our capacity to create.

The process of learning is composed of several activities, among which the most important are perception, knowledge acquisition, knowledge structuring, and restructuring through a continuous dynamics, knowledge storage, knowledge removal from memory, and knowledge creation through a conscious effort. In his research about learning organizations, Bob Garratt demonstrates that "organizations can only become simultaneously effective and efficient if there is conscious and continuous learning between three distinct groups—the leaders who direct the enterprise, the staff who deliver the product or service, and the customers or consumers" (Garratt, 2001, p. IX).

The organizational type of learning differs from individual learning, where there is only one cycle going from practice to conceptualization and testing, through tacit and explicit knowledge. (Organizational learning contains three main cycles of learning: the operational learning cycle, the strategic learning cycle, and the policy learning cycle.)

- The *operational learning cycle is a component of the operational management* (based on the daily activities, the timescale ranging at the most up to 1 year, operational management is concerned mostly with the process of production and its efficiency, with economic rationality and short-term objectives, the most obvious characteristics of the operational management; the operational learning cycle produces innovation at execution line, both technological and managerial).
- *The strategic learning cycle refers to the bridging together the policy learning cycle and the operational learning cycle* (strategic learning is about monitoring the changing external world, reviewing the organization's position in these changes, making risk assessments to protect and develop enterprise, broadly deploying its scarce resources to achieve its purpose, and ensuring that there are feedback procedures in place to measure the effectiveness of any strategy being implemented).
- The *policy learning cycle contains the organization relationships with the external business environment* (controlled from inside by top management, and people in charge with establishing a company's policy, that must understand the complexity of the new unpredictable and chaotic external business environment; experience shows that the customer's or consumer's perception of organization effectiveness contributes directly to the success or failure of that organization).

In the university internal environment, the process of production is a *knowledge generation* and *transfer process*, and the process of management deals also with knowledge. Thus, both processes perform in the world of intangibles, and the production process is actually limited by the performance capacity of the management process. In this context, the paradox may have a sense since the production process is actually a learning process. If a certain university is going to be a *learning organization*, it is necessary that the process of management should become a learning process, as well.

As Chris Argyris demonstrates, "Learning occurs when the invented solution is actually produced. This distinction is important because it implies that discovering problems and inventing solutions are necessary, but not sufficient conditions, for organizational learning. Organizations exist in order to act and to accomplish their intended consequences." (Argyris 1999, pp. 68–69).

In the dynamic process of transformation of individual contributions of all the organization members into the organizational entities, in terms of knowledge, intelligence and values, the major role is played by *integrators* (according to Constantin Brătianu). The *team management* acts as an integrator at the team level: "*an integrator is a powerful field of forces capable of combining two or more elements into a new entity, based on interdependence and synergy. These elements may have a physical or virtual nature, and they must possess the capacity of interacting in a controlled way*" (Brătianu, 2007b, p. 111).

Management is by its own nature an integrator, sometimes equal to, but often more powerful than technology and its associated fundamental sciences. The technology integrator or new important scientific disciplines are capable of acting only upon the explicit knowledge, which is codified in a certain way. The management integrator can act upon both explicit and tacit knowledge, generating explicit organizational knowledge and tacit organizational knowledge.

This could be econophysics' third argument for its models and methods, which are really the best solutions in the new era of IT, and its information containing more than 10^{23} data. Econophysics is probably the best integrator in a modern university.

10.6 CONCLUSIONS

In a comparison with classical sociology, the new science of sociophysics has revealed that the heterogeneous in social reality must be explained with the homogeneous in theory, and this is the most important improvement of the quality in classical science and research. The main role of physics and its methods, like statistical physics or quantum physics for the beginning, was to unify and simplify classical sociology. Scientific sociophysics research improves the quality of classical sociology and physics, and extends their themes, fields, models, and interpretations. Some new contradictions have appeared in the new science, named simply sociophysics, dealing with the following:

- An interesting niche in computer research, where it has been established by making models much simpler than most economists or sociologists now choose to consider even using possible connection between sociologic or economic terms and *critical points* in statistical mechanics (of course, one needs to be careful about analogies and model simplifications; many of these models are heuristic, they can help us in understanding principles and do not necessarily describe the complexity of individual economic and social cases).
- A response of a physical system to a small external perturbation, which becomes infinite because all the subparts of the system respond cooperatively, or the concept of *noise* in spite of the fact that some economists and sociologists even claim that it is an insult to the intelligence of the market or of the society to invoke the presence of a *noise* term. The power of prediction or the higher level of accuracy of econophysics and sociophysics models remain the most important difference between these new and powerful methods and other sciences and classical sociology thinking. These models are often better than classical econometrical or statistical correlation models. The complexity studies of sociophysics try to capture the universal but temporary laws, from data manifested differently in different parts of the same body of natural phenomena, where information about population is made from more individuals than 10^{23} units (cases). This grand unification search has reached a really inspiring stage today, and the present contribution reports on a part of these interdisciplinary studies, developed over the last 20−25 years, and classified under the heading of sociophysics.

It was not only the remarkable results of these new sciences that motivated us to collect these authentic reviews on intriguing qualitative developments of sociophysics but also the rapid success in solving difficulties in the social and economic contemporary reality and the way in which this science has improved the quality of classical science sociology. Instead of the promise and novelty of these new researches, it was the curiosity to understand how a new science has solved the problems which has been a guide in selecting articles and books, techniques and methods, models, and temporary laws. The future scientific thought will be nothing else but statistical; it will be either a generalized thinking as in statistical physics or a classical distinctive statistical thinking.

From a sociophysicist's point of view, the individuals, their behavior, and the interactions between all of them constitute a microscopic level for any social system (from a family to a generalized concept, as public opinion is). The major question remains the same: Can the laws on the microscopic scale of a social system explain phenomena on the macroscopic scale (resulting in a higher R^2), that sociologists deal with?

10.7 A FINAL REMARK

This chapter was devoted to the cross-fertilization of interdisciplinary fields within sociophysics. It seems possible in the future for the boundaries between sciences to be considered more as determined by methods and not by the subjects of research. Over the last two decades, sociophysics has increasingly established itself. The most important problem for this new science certainly is the ability to understand the rapid change in the realities of economic and social life. It will be neither the strongest science nor the most methodological that survives; it will be the one that can adapt itself to changes most rapidly and frequently, finding the best methods, techniques, instruments, concepts, and solutions. This must be the science with the way of thinking best suited to reality; it will be one of the most necessary disciplines in a modern university.

REFERENCES

Albrecht, K., 2003. The Power of Minds at Work: Organizational Intelligence in Action. American Management Association, New York, NY.

Argyris, C., 1999. On Organizational Learning, 2nd edn. Blackwell Business, Oxford.

Arnopoulos, P., 1993. Sociophysics: Chaos and Cosmos in Nature and Culture. Science Publishers, New York.

Ball, P., 2004. Critical Mass: How One Thing Leads to Another. Arrow Books, London.

Bernardes, A.T., Costa, U.M.S., Araujo, A.D., Stauffer, D., 2001. Damage spreading, coarsening dynamics and distribution of political votes in Sznajd model on square lattice. Int. J. Mod. Phys. C 12 (2), 159−167.

Brătianu, C., 2007a. The learning paradox and the university. J. Appl. Quant. Methods 2 (4), 375−386.

Brătianu, C., 2007b. An integrative perspective on the organizational intellectual capital. Rev. Manage. Econ. Eng. 6, 107−113.

Chakraborti, A., 2008. Methods in econophysics: successes and failures. Brochure of International Workshop and Conference: Statistical Physics Approaches to Multi-disciplinary Problems, 7−13 January, IIT, India, pp. 3 − 4.

De Oliveira, M.S., de Oliveira, P.M.C., Stauffer, D., 1999. Evolution, Money, War, and Computers: Non-Traditional Applications of Computational Statistical Physics. Teubner, Stuttgart-Leipzig.

Deffuant, G., 2002. How can extremism prevail? A study based on the relative agreement interaction model. J. Artif. Soc. Social. Simulat. 5 (4), paragraph 1−7. <http:/jasss.soc.surrey.ac.uk>.

Galam, S., 1984. Entropy. Semiotext(e) 4, 73−74.

Galam, S., 1986. Majority rule, hierarchical structures and democratic totalitarianism: a statistical approach. J. Math. Psychol. 30, 426−434.

Galam, S., 1990. Social paradoxes of majority rule voting and renormalization group. J. Stat. Phys. 61, 943−951.

Galam, S., 1991. Political paradoxes of majority rule voting and hierarchical systems. Int. J. Gen. Syst. 18, 191−200.

Galam, S., 1999. Application of statistical physics to politics. Phys. A: Stat. Mech. Appl. 274, 132−139.

Galam, S., 2002. Minority opinion spreading in random geometry. Eur. Phys. J. B 25, 403−406.

Galam, S., 2003. Modeling rumors: the no plane pentagon French hoax case. Phys. A: Stat. Mech. Appl. 320, 571−580.

Galam, S., 2004. Sociophysics: a personal testimony. Phys. A: Stat. Mech. Appl. 336, 49−55.

Galam, S., 2008a. Sociophysics: a review of Galam models. Int. J. Mod. Phys. C 19, 409−440.

Galam, S., 2008b. La Science Magique et le Rechauffement du Climat. Editions Plon, Paris.

Galam, S., 2011. Collective beliefs versus individual inflexibility: the unavoidable biases of a public debate. Phys. A: Stat. Mech. Appl. 390 (17), 3036−3054.

Galam, S., Gefen, Y., Shapir, Y., 1982. Sociophysics: a mean behavior model for the process of strike. J. Math. Sociol. 9, 1−13.

Galam, S., Jacobs, F., 2007. The role of inflexible minorities in the breaking of democratic opinion dynamics. Phys. A: Stat. Mech. Appl. 381, 366−376.

Galam, S., Moscovici, S., 1991. Towards a theory of collective phenomena: consensus and attitude changes in groups. Eur. J. Social Psychol. 21, 49−74.

Garratt, B., 2001. The Learning Organization: Developing Democracy at Work. Harper Collins Business, London.

Helbing, D., 1995. Quantitative Sociodynamics: Stochastic Methods and Models of Social Interaction Processes. Kluwer, Dordrecht, The Netherlands.

Helbing, D., 2010. Quantitative Sociodynamics: Stochastic Methods and Models of Social Interaction Processes, 2nd edn. Springer, The Netherlands.

Hu, C., Wu, R., Liu, J.Y., 2012. Study on the model of consensus formation in Internet based on the directed graph. Int. J. Mod. Phys. C 23 (6), doi: 10.1142/S0129183112500416.

Klemm, K., Eguıluz, V.M., Toral, R., San Miguel, M, 2005. Globalization, polarization and cultural drift. J. Econ. Dyn. Control 29 (1 − 2), 321−334.

Lind, P.G., Andrade Jr., J.S., da Silva, L.R., Herrmann, H.J., 2007. Spreading gossip in social networks. Phys. Rev. E 76, 036117.

Mantegna, R.N., Stanley, H.E., 2000. An Introduction to Econophysics: Correlations and Complexity in Finance. Cambridge University Press, Cambridge.

Robbins, S.P., DeCenzo, D.A., 2005. Fundamentals of management. Essential concepts and applications, 5th edn Pearson, Prentice Hall, London.

Rosser Jr., J.B., 2006. The nature and future of econophysics. In: Chatterjee, A, Chakrabarti, B.K (Eds.), Econophysics of Stock and Other Markets. Springer, Milan, pp. 225−239.

Schulze, C, Stauffer, D., 2006. Recent developments in computer simulations of language competition. Comp. Sci. Eng. 8, 60−67.

Senge, P.M., 1990. The Fifth Discipline: The Art and Practice of the Learning Organization. Random House, London.

Sobkowicz, P., 2011. Simulations of opinion changes in scientific communities. Scientometrics 87 (2), 233−250.

Stauffer, D., 2003. Sociophysics simulations. Comput. Sci. Eng. 5 (3), 71−75.

Stauffer, D., 2005. Sociophysics simulations IV: Hierarchies of Bonabeau et al. AIP Conf. Proc. 779, 75−80.

Stauffer, D., de Oliveira, M.S., de Oliveira, P.M.C., San Martins, J.S., 2006. Biology, Sociology, Geology by Computational Physicists. Elsevier, Amsterdam, The Netherlands.

Sznajd-Weron, K., 2004. Dynamical model of Ising spins. Phys. Rev. E 70, 037104.

Sznajd-Weron, K., 2005. Sznajd model and its applications. Acta Phys. Pol. B 36 (8), 2537−2547.

Sznajd-Weron, K., Pękalski, A., 2001. Model of population migration in a changing habitat. Phys. A: Stat. Mech. Appl. 294, 424.

Sznajd-Weron, K., Sznajd, J., 2000. Opinion evolution in closed community. Int. J. Mod. Phys. C 11 (6), 1157−1165.

Vitanov, N.K., Dimitrova, Z.I., Ausloos, M., 2010. Verhulst−Lotka−Volterra (VLV) model of ideological struggle. Phys. A: Stat. Mech. Appl. 389 (21), 4970−4980.

Vitanov, N.K., Dimitrova, Z.I., Stojcho Panchev, S., 2005. Challenges to contemporary physics: econophysics and sociophysics. Science (Sofia) 15, 13−23.

Weidlich, W., 2000. Sociodynamics: A Systematic Approach to Mathematical Modelling in the Social Sciences. Harwood Academic Publishers, Amsterdam, The Netherlands.

Printed in the United States
By Bookmasters